T0347376

Tunisia

The ILO's World Employment Programme (WEP) aims to assist and encourage member States to adopt and implement active policies and projects designed to promote full, productive and freely-chosen employment and to reduce poverty. Through its action-oriented research, technical advisory services, national projects and the work of its four regional employment teams in Africa, Asia and Latin America, the WEP pays special attention to the longer-term development problems of rural areas where the vast majority of poor and underemployed people still live, and to the rapidly growing urban informal sector.

At the same time, in response to the economic crises and the growth in open unemployment of the 1980s, the WEP has entered into an ongoing dialogue with the social partners and other international agencies on the social dimensions of adjustment, and is devoting a major part of its policy analysis and advice to achieving greater equity in structural adjustment programmes. Employment and poverty monitoring, direct employment creation and income generation for vulnerable groups, linkages between macro-economic and micro-economic interventions, technological change and labour market problems and policies are among the areas covered.

Through these overall activities, the ILO has been able to help national decision-makers to reshape their policies and plans with the aim of eradicating mass poverty and promoting productive employment.

This publication is the outcome of a WEP project.

Tunisia
Rural labour and structural transformation

Samir Radwan, Vali Jamal

and

Ajit Ghose

A study prepared for the International Labour Office within the framework of the World Employment Programme

London and New York

First published 1991
by Routledge
2 Park Square, Milton Park, Abingdon, Oxon, OX14 4RN

Simultaneously published in the USA and Canada
by Routledge
270 Madison Ave, New York NY 10016

Transferred to Digital Printing 2007

© 1991 International Labour Organisation

Phototypeset in 10pt Times by
Mews Photosetting, Beckenham, Kent

British Library Cataloguing in Publication Data

Radwan, Samir
 Tunisia: Rural labour and structural transformation.
 1. Tunisia. Economic conditions
 I. Title II. Jamal, Vali III. Ghose, Ajit Kumar,
 330.9611052

 ISBN 0-415-04274-7

Library of Congress Cataloging in Publication Data
has been applied for
 ISBN 0-415-04274-7

Cover design: Teresa Dearlove

Publisher's Note
The publisher has gone to great lengths to ensure the quality of this reprint but points out that some imperfections in the original may be apparent

Contents

List of figures   vii

List of tables   viii

Preface   xi

1   **Introduction**   1

2   **Employment and labour markets**   8
    Trends in population and labour force   8
    Agricultural labour force   11
    Traditional escape routes   17
    Conclusion   27

3   **The agrarian economy: Evolution and prospects**   29
    Evolution of agrarian policy   30
    Trends in agricultural production   31
    Technological change   35
    Some aspects of landholding   36
    Investment in agriculture   39
    Prices and subsidies   41
    Conclusion   44

4   **Income distribution and poverty**   45
    Income distribution and trends   45
    Poverty incidence and trends   52
    Growth and equity   61
    Conclusion   61

5   **Food consumption, food balances and subsidies**   63
    Food consumption in 1980   63
    Cereal preferences   66
    Food balances   74

Contents

	Subsidies	82
	Conclusion	85
6	**Conclusion**	87
	Appendix A: Labour migration from Tunisia	92
	Appendix B: Estimates of unemployment	95
	Appendix C: Poverty lines in Tunisia: A critical review	98
	Notes	113
	Index	118

Figures

4.1	SMIG in real terms, 1961–87	60
5.1	Food consumption–income relationships, 1980	66
5.2	Wheat preferences by income classes	68
5.3	Cereal preferences: rural and urban, 1980	70
5.4	Hard and soft wheat: quantity and price ratios, 1966–80, and quantity ratio projected to 2000	73
5.5	Wheat production, imports and total cereal production, 1934–38, 1948–52 and 1952–87	79
5.6	Food balances, 1934–38, 1962–64 and 1982–84	81

Tables

1.1 External resources and GDP,1965–86   4
1.2 Vital indicators of the economy, 1965–86   5
2.1 Distribution of population and labour force by location, 1956–84   9
2.2 Labour force by sectors, 1975 and 1984   12
2.3 Structure of the labour force by location, 1984   13
2.4 Agricultural labour force by type, 1976/77 to 1984/85   14
2.5 Employment by size of landholding, 1979   15
2.6 Non-agricultural activities, 1985   16
2.7 Distribution of landholders by time devoted to agricultural activities and size of holding, 1980   17
2.8 Labour force breakdown by major forms, 1975, 1980 and 1984   18
2.9 Informal sector employment by localization compared with public and private modern sector employment, 1980   20
2.10 Ratio of value added per capita in informal sector compared with modern sector, 1981   21
2.11 Net emigration and remittances, 1964–86   24
2.12 Contribution of emigration to employment, 1962–81 and projections for 1982–91   25
2.13 Regional characteristics of active population and unemployment, 1975 and 1984   27
3.1 Average annual rates of growth of agricultural output, 1962–64 to 1982–84, selected years   32
3.2 Cropping pattern, 1962 to 1982–84, selected years   33
3.3 Allocation of irrigated area, 1983–84   34
3.4 Livestock and poultry population, 1956 to 1982–84, selected years   34
3.5 Indicators of technological change, 1960–62 to 1982–84   35

3.6 Pattern of landholding, 1961-62, 1975-76 and
1979-80   37
3.7 Distribution of cultivated areas, 1975-76 (annual
crops and tree crops) and 1979-80 (annual crops)   37
3.8 Pattern of fertilizer use by size-class of holding,
1975-76 and 1979-80   38
3.9 Livestock distribution, 1975-76 and 1979-80   39
3.10 Investment in agriculture, 1960-61 to 1977-84   40
3.11 Percentage distribution of investment in agriculture by
subsector, 1962-71 to 1982-85   40
3.12 Growth of product prices, 1965-1969/71 to
1976/78-1982/84   42
4.1 Expenditure per person and per household, 1980 and
1985   46
4.2 Per capita consumption by region, 1980 and 1985   47
4.3 Consumption according to social classes, 1980 and
1985   47
4.4 Consumption distribution, 1985   48
4.5 Consumption distribution according to household
budget surveys   49
4.6 Agricultural and non-agricultural GDP and wage
share, 1966-84, selected items and years   51
4.7 Poverty lines and poverty incidence, 1975 and 1985:
ILO, World Bank and INS   53
4.8 Incidence of poverty and total number in poverty by
occupational groups, 1985   54
4.9 Incidence of poverty by districts and rural/urban,
1985   55
4.10 Poverty line and poverty incidence, 1966-85   57
4.11 Total number in poverty: ILO and INS, 1975, 1980
and 1985   57
4.12 SMIG (48-hour week) in current and real terms,
1961-87   59
5.1 Calorie and expenditure distribution of the average
diet, 1980   64
5.2 Food consumption-income relationships, 1980   65
5.3 Wheat preferences: hard and soft wheat, 1980   67
5.4 Cereal preferences: rural and urban, 1980   69
5.5 Changing wheat preferences, 1966-85   69
5.6 Price regime: hard and soft wheat, 1980 and 1987   72
5.7 Evolution of trade balance, selected items, 1975,
1980 and 1985   74
5.8 Food imports, 1983-85   75
5.9 Cereal supply, 1977-86, selected years   76

Tables

5.10	Wheat demand and supply, 1980	77
5.11	Production and imports of wheat and total cereals, 1934–38 to 1985 selected years	78
5.12	Estimated food balance sheets for 1934–38, 1962–64 and 1982–84	80
5.13	Expenditures of the price equalization fund, 1974–83	83
5.14	Subsidies and price support in Tunisia, 1982	84
5.15	Food subsidies in Tunisia and neighbouring countries, 1982	85
A.1	Estimate of Tunisian migrants in Europe, 1984	93
A.2	Estimate of Tunisian migrants in Arab countries, c. 1985	93
B.1	Structure of the labour force, 1975	96
B.2	Structure of the labour force, 1984	96
C.1	Absolute poverty threshold according to the World Bank	100
C.2	Cost of a cereal diet, 1966	103
C.3	Poverty lines and poverty in four Development Plans	105
C.4	Composition and cost of a minimum food basket, 1975	108
C.5	Cost of three different diets at 1975 prices	109
C.6	Prices in 1966 and 1980	109
C.7	Average expenditure from household budget surveys and estimated poverty lines	111
C.8	Cost of a food basket in 1985	112

Preface

This study provides a contribution to the analysis of structural transformation in Tunisia over the last two decades. During this period, the country had a spectacular growth in comparison with many developing countries. Per capita income increased at 5 per cent per annum, reaching a level of $1,300 by the early 1980s, propelling the country to the ranks of 'middle-income' countries. Growth was accompanied by an equally spectacular transformation of the economy whereby the labour force profile changed from a predominantly agricultural to a predominantly urban orientation. The adjustments in the labour markets, the changes in the agricultural sector which facilitated these, and the implications of the growth and structural transformation for equity and food balances are the issues addressed in this volume.

The study represents the culmination of a collaborative effort between Tunisian researchers, the League of Arab States and the ILO. A number of background studies[1] were undertaken by prominent Tunisian researchers drawn from the Commissariat Général de Développement Régional of the Ministry of Planning, the Ministry of Agriculture, and the University. The analysis presented here draws on these background studies as well as on data gathered by the authors. Preliminary findings were presented at a national seminar in Tunisia in May 1987. As a follow-up, a project was launched in 1988 by the Ministry of Social Affairs under the title of 'Lutte contre la pauvreté' to assist in identifying poor families.

In writing this study, Samir Radwan was responsible for the overall design and principal theme, and contributed Chapter two; Vali Jamal wrote Chapters four and five; and Ajit Ghose wrote Chapter three. Jamal and Radwan were responsible for the editing of successive versions of the manuscript and its preparation for publication. The responsibility for the final book is, however, collectively shared.

The completion of this study would have been impossible without the help and co-operation of many people in Tunisia. It would be difficult to provide the full list of those people. However, mention must be made of a few. Our foremost debt is to Dr. Darim al Bassam, Director of the

xi

Population Research Unit of the Arab League who took a personal interest in all stages of our work, provided constant encouragement, and ensured the co-operation of his colleagues, Hafez Schikir and Khaled Louchici. Bedoui Abdel-Jelil, Mohamed Ayyad, Messaoud Boudiaf, Hussein Dimacy, Omar Kaddour, Khalil Zamiti and Mongi Bougazala were our Tunisian research counterparts and contributed to our knowledge of the economy through continual formal and informal encounters and background papers. At the government level our thanks go to the Minister of Social Affairs M. Hedi Baccouche for his encouragement, to Messrs Abdelmajid Mabrouk, Ali Sanaa and Hajh Dahmane of his Ministry for their close support, and to Mr Abdelssalem Kammoun, the ex-Director General of the Institut National de la Statistique, and his staff for so willingly sharing with us their published and unpublished data.

At the ILO we are pleased to record our thanks to Mr Jack Martin, the then Director of the Employment Department under whom the project was initiated and executed, to Victor Tokman, Rolph van der Hoeven, Michael Hopkins and Hamid Tabatabai for their constructive comments, and to Geraldine Ople and Cheryl Wright for cheerfully typing the various drafts. Needless to say the responsibility remains that of the authors.

It is hoped that the study provides an objective analysis of the problems facing Tunisia as it tries to forge a new direction for the future.

Samir Radwan
Chief
Rural Employment Policies Branch
Employment and Development Department, ILO

Chapter one

Introduction

Up to the early 1980s Tunisia was considered a model of successful development. With a per capita income of US$1,300 in 1980, it had joined the middle ranks of middle-income countries, in World Bank terminology, in company with Turkey, Jamaica, Guatemala and so on. Eighteen years earlier its per capita income was only around US$500. The implied rate of growth of 5 per cent per annum achieved during this period was equalled or exceeded by only ten other countries among the 126 listed by the World Bank. By the mid-1980s, however, the material conditions facing Tunisia had changed drastically. Growth had slowed, the balance of payments had deteriorated and unemployment had begun to increase. These trends continued into the late 1980s despite a resurgence of tourism and bountiful harvests. 'Crisis' became the commonest descriptor of the economy, as debate increased on the structural problems of the Tunisian economy. Among the issues that figured in this debate, five in particular stood out: (1) the nature of the structural transformation of the economy; (2) changes in the labour force; (3) problems of the agrarian economy; (4) trends in income distribution and poverty; and (5) the state of food balances. These issues of course occupy the centre stage in any discussion of economic development. What makes them so relevant in the Tunisian case is that Tunisia achieved a unique transformation of its economy within a short period, altering its predominantly rural character to one that is predominantly urban. What sort of shift in the labour force did this imply? What was the role of the agricultural sector in this shift and what was the outcome for equity and poverty? These are the major questions that have to be addressed to evaluate the overall performance of the economy. Although we focus on a specific country we do believe our analysis has lessons of general applicability, particularly since, apart from the newly industrializing countries of East Asia, not many developing countries have exhibited such a rapid transformation of their economies.

The debate in Tunisia at mid-1980 centred around the character and pace of structural transformation undergone by the economy in the previous two decades. Three different perceptions could be distinguished. Each, it will be noticed, had aspects of the other two, so that in a sense they were more complementary than competitive. The first perspective best represented by the government, argued that the major problem to be tackled by the country was unemployment. The Sixth Development Plan (1982–86) was blamed for failing to generate sufficient employment because of an over-emphasis on capital-intensive projects. External factors, especially increasing debt service, falling terms of trade and the international recession, exacerbated the financial crisis. The next development plan (the Seventh Plan, 1986–90) thus proposed an export-oriented growth strategy as the solution. The second perspective, represented by the World Bank and the International Monetary Fund (IMF), viewed things differently. The 1970s witnessed an externally generated boom through an increase in oil exports, tourism and remittances of migrant workers. The markets functioned well. However, by the early 1980s the government's programme of expenditure led to the accumulation of debt, a deficit in the balance of payments and a decline in the external resources. The situation reached crisis proportions in the mid-1980s with the drastic fall in the price of oil. Thus what was required was a 'stabilization programme', the main features of which were currency devaluation, price liberalization, removal of subsidies and a reduction in public expenditure. The third view of the economy was one usually held by Tunisian economists and trade unionists. This perceived the crisis in a historical perspective in which the problem was diagnosed as structural rather than conjunctural, with the blame being laid on a pattern of development characterized by dependency. It was argued that both the government and the World Bank solutions would exacerbate the crisis as they did not visualize any basic alteration in the relationship of the economy with the outside world or in the overall development strategy.

This study starts from the premiss that the crisis has to be seen in terms of the action of the state in the acquisition and distribution of 'rent' from external stimuli. In the 1970s the country enjoyed a favourable foreign exchange position due to a rapid growth in tourism, petroleum exports and remittances. The state became the supreme intermediary of the rent arising from these sources. The development strategy followed was characterized by heavily subsidized investment financed from that rent. Part of these subsidies went into the creation of an industrial structure protected behind high tariff walls, and the other part into the agricultural sector in the form of inputs and services. The ensuing 'pull and push' factors engendered a massive transformation of the economy with a significant turnaround in the roles of agriculture and non-agriculture as repositories of the labour force. This structural transformation was dependent largely on the infusion of externally generated rental incomes into the economy. Its beneficiaries were highly subsidized and dependent agricultural and industrial sectors whose basic deficiencies were masked by the boom. When the boom collapsed, the stage was set for the revelation of the crisis in all its dimensions.

While the boom period intensified the role of the state in shaping development in Tunisia, such a role has been its prerogative ever since independence (1956). Three phases can be distinguished:

1 1956–62: the period of decolonization. The state became the repository of windfall gains as a result of the departure of foreigners living in Tunisia. By some estimates, around 200,000 people left the country during this period. Apart from French and Algerian nationals these also included Libyans who used to live and work in Tunisia and who were attracted back to their native land by the discovery of oil. The departure of the 'colons' freed vast areas of land in the best arable zones, which were turned into state domains or to a limited extent, distributed to individual farmers. This fortuitous distribution of land – limited though it was – obviated the need to carry out land reform to satisfy the rural population. In the urban areas too the employment problem was eased because of the vast number of jobs vacated by migrants that could be filled by nationals. After this initial distribution of employment opportunities the economy stagnated and underemployment began to increase in the rural areas.

2 1962–72: the decade of the agricultural co-operatives. An ambitious programme of institutional transformation was introduced, with the objective of organizing individual farmers in agricultural co-operatives and thus facilitating the mechanization and modernization of traditional agriculture. This experiment has given rise, until the present, to heated controversy as to the pros and cons of this forced transformation of agricultural structures. A consensus, however, does exist on two points: (1) the main source of financing development during this period was the transfer of agricultural surplus as represented by exports and the supply of cheap food and raw materials to the urban areas; and (2) there was a concomitant exodus of labour from the rural areas to urban centres and abroad, especially to France.

3 1972–82: the boom period. Its major characteristic was an increase in resources at the disposal of the government from petroleum, remittances and tourism. A large part of this rent was transferred to agriculture and industry. Tunisian agriculture witnessed a highly subsidized programme of investment and the provision of cheap credit. Similarly, a process of subsidized industrialization began in the processing and import-substituting sectors. At a later stage, a number of industries producing for export were established, especially in the textile sector. These export-oriented industries remained dependent

on imported inputs. The decade also witnessed the beginning of a process of migration to the Libyan Arab Jamahiriya, which eventually replaced France as the main foreign outlet for Tunisian labour.

As is evident from the above description of government policies, the driving force in the Tunisian economy in the 1970s was provided by the infusion of foreign exchange. Table 1.1 presents some relevant data on this in terms of the three main components of foreign exchange: remittance, tourism receipts and petroleum revenues. In 1965 the contribution of remittances to gross domestic product (GDP) was negligible; by the late 1970s this contribution stood at nearly one-quarter of the GDP. Although exact figures for earlier years are not available, we would estimate that foreign receipts increased at least fifty-fold between 1965 and 1977.

Table 1.1 External resources and GDP, 1965–86 (million dinars and percentages)

Year	Remittances	Tourism	Crude petroleum exports	Total	As % of GDP	Memo item: GDP at current prices
1965	3.8	n.a.	n.a.	n.a.	n.a.	548.0
1975	58.2	n.a.	n.a.	n.a.	n.a.	1,723.0
1976	61.4	n.a.	139.0	n.a.	n.a.	1,908.0
1977	72.2	135.0	161.0	368.2	16.8	2,188.0
1978	91.8	165.0	180.0	436.8	17.6	2,482.0
1979	115.5	219.0	327.0	661.5	23.0	2,878.0
1980	129.2	259.6	450.0	838.8	24.0	3,510.0
1981	178.3	295.2	625.0	1,098.5	26.4	4,162.0
1982	219.6	340.0	507.0	1,066.6	22.0	4,817.0
1983	235.0	389.0	536.0	1,160.0	21.0	5,520.0
1984	246.0	357.0	581.0	1,184.0	19.0	6,235.0
1985	210.0	n.a.	n.a.	n.a.	n.a.	n.a.
1986	190.0 (est.)	n.a.	n.a.	n.a.	n.a.	n.a.

Sources: Annuaire statistique de la Tunisie, several issues; Banque Centrale de Tunisie, *Rapport annuel*, several issues.
Note: n.a. = not available

Some indicators of the ensuing growth are shown in Table 1.2. Between 1965 and 1973 GDP increased at 7.3 per cent per annum, while in the next ten years it increased at 6.0 per cent per annum. Growth was more or less equally shared among the sectors in the first period but in the second period manufacturing growth far outstripped agriculture. A transformation of the economy was effected, with agriculture's share of the labour force falling quite perceptibly, from one-half in 1965 to just around one-third 15 years later. Agricultural labour force in the

Table 1.2 Vital indicators of the economy, 1965–86

Positive				
GNP per capita, 1986		$1.140		
Growth rate per annum (%)	1965–83	5.0		
	1965–86	3.8		
		GDP	*Agriculture*	*Manufacturing*
Growth rate (% p.a.)	1965–83	6.6	5.5	9.9
	1965–73	7.3	6.9	10.3
	1973–83	6.0	1.6	11.1
	1980–86	3.7	3.3	6.5
		Agriculture	*Industry*	*Services*
Labour force (%)	1965	49	21	29
	1980	35	36	29
	(1986)	(22)	n.a.	n.a.
Daily calorie supply per capita	1965	2,296		
	1985	2,796		
Percentage of requirement		122		
Population per physician	1965	8,000		
	1981	3,620		
Infant mortality rate	1965	145		
	1986	74		
		Male	*Female*	
Life expectancy	1965	51	52	
	1986	61	65	
Negative				
GNP per capita, growth rate per annum, 1983–86		−3.1		
Cereal imports ('000 tonnes)	1974	307		
	1986	1,312		
Current account balance (m$)	1970	−53		
	1986	−657		
		% GNP	*% Exports*	
Debt service	1970	4.7	19.7	
	1986	10.0	30.8	

Sources: IBRD (World Bank), *World Development Report, 1988*, various tables, except growth rate of GDP, etc. which is from *World Development Report, 1985*, and labour force in 1986 which is extracted from Table 2.1.

aggregate nonetheless grew by about one-fifth, as compared with an increase of over 2.5 times in the non-agricultural labour force.

Undoubtedly this structural transformation had many positive aspects. Non-agricultural employment grew at nearly 3 per cent per annum between 1975 and 1984 and all sectors shared in the growth, although with the informal sector dominating employment, a great part of the increasing non-agricultural labour force was absorbed there. Growth was also associated with improvements in real welfare. The average daily supply of calories easily surpassed the requirement (around 2,300 calories) from a level just below requirement, while the number of doctors per inhabitant increased appreciably. The consequence was that infant mortality declined by almost half, while life expectancy increased from around 52 to 63 years. This last statistic is increasingly regarded as the best indicator of real welfare, since it embodies underlying changes in health, nutrition and associated variables, and since it reflects essentially long-term trends. On this basis it could be said that the quality of life of the Tunisian people improved significantly during the two decades of economic growth. There are instances – oil-boom countries being prime examples – where increasing incomes leave all other social variables behind: the country joins the ranks of middle- or even high-income countries in terms of income, but stays with the low-income countries in terms of social amenities. Tunisia can rightfully claim its place among the countries in which its per capita income places it.[1]

However, negative signs began to surface in the early 1980s. Aggregate GDP stagnated between 1983 and 1986, so that per capita GDP fell by some 9 per cent. Agriculture's contribution to the slow-down was minimal or even zero, as agricultural output per capita actually increased by 6.2 per cent during the same period. Agriculture had in fact failed before this. Thus, between 1975 and 1983 agriculture output per capita declined by 16 per cent. However, agricultural productivity increased slightly because of the great rural-to-urban migration that was going on throughout this period. As the last two statistics show, the Tunisian 'crisis' took a very conventional form. The current account imbalance, which had always been a feature of the economy – even at the height of the oil-price boom – worsened considerably in the 1980s. A related consequence was that the debt service ratio increased from 4.7 per cent of GNP in 1970 to 10 per cent in 1986, or from 20 per cent to 31 per cent in terms of exports.

From this preliminary assessment of macroeconomic magnitudes, many positive aspects of the Tunisian experience come through, particularly in terms of common indicators such as economic growth, structural transformation and improvement in living standards. Concern may exist at the crisis hanging over the economy but the crisis can be differentiated from past achievements. Three questions still remain to be answered: Was the structural transformation genuine – that is, did it occur because of balanced growth in the agricultural and non-agricultural sectors? Were the structures created strong enough to sustain self-generated growth in the economy? Did all sectors of the population share in the growth? These broad themes constitute the subject matter of this monograph. The first two are addressed in Chapters 2 and 3, and in terms of food balances, in Chapter 5. The third question is discussed in Chapter 4, while the concluding chapter brings the various strands of the analysis together to give our final assessment of the economy.

Chapter two

Employment and labour markets

The creation of employment has been a preoccupation of the government since independence. While the country has succeeded – with the help of external booms – in achieving growth, it has been much less successful in reaching its employment objective. Successive waves of external migration and spurts of rural-urban exodus indicate that the agrarian economy was 'pushing out' increasing numbers of the labour force. Development plans as well as projects and programmes were promulgated to assist the migrants in the urban areas and the coastal tourist belt. Yet chronic unemployment, especially in the disadvantaged regions of the north-west and the south, posed an even greater challenge. The central question thus remains of how the employment problem is to be solved. How and to what extent was the increase in the labour force absorbed in agricultural or non-agricultural activities? The answer to these questions is best provided in terms of the structural transformation undergone by the economy in the last three decades. In this chapter we look at the mechanics of this transformation, focusing on population movements. Our concern will be to establish where the transferring population went, and particularly whether genuine expansion occurred in the non-agricultural sectors to accommodate the growing labour force.

Trends in population and labour force

Between 1956 and 1984 Tunisia's population increased at 2.2 per cent per annum to register a total growth of 84 per cent (Table 2.1), reaching almost 7 million inhabitants. The inter-censal data indicate that growth accelerated throughout this period, starting at 1.8 per cent for the period 1956–66 and reaching 2.5 per cent between 1975 and 1984. This acceleration was due to falling infant and total mortality rates contingent upon improving health standards in the country. Thus, between 1965 and 1984 the infant mortality rate of children (under one year of age) nearly halved, the child death rate fell by three-quarters and the crude death rate was cut by half (from 18 per thousand inhabitants to 9).

Table 2.1 Distribution of population and labour force by location, 1956–84 (thousands and percentages)

	1956		1966		1975		1984		Annual rate of growth (in per cent)			
	Number	Per cent	Number	Per cent	Number	Per cent	Number	Per cent	1956–66	1966–75	1975–84	1956–84
Population												
Urban	1,252	33.1	1,820	40.2	2,779.0	49.8	3,681.0	52.8	3.8	4.8	3.2	3.9
Rural	2,531	66.8	2,713	59.8	2,798.0	50.2	3,285.0	47.2	0.7	0.3	1.8	1.0
Total	3,783	100.0	4,533	100.0	5,577.0	100.0	6,966.0	100.0	1.8	2.3	2.5	2.2
Labour force												
Urban	n.a.	n.a.	464.8	42.5	828.9	51.1	1,166.7	54.4	n.a.	6.6	3.9	
Rural	n.a.	n.a.	628.9	57.5	792.9	48.9	970.4	45.4	n.a.	2.6	2.2	
(Agricultural)					(509)	(31.4)	(475)	(22.2)				
Total	n.a.	n.a.	1,093.7	100.0	1,621.8	100.0	2,137.2	100.0	n.a.	4.5	3.1	

Source: Institut National de la Statistique (INS). *Recensement de la population 1956, 1966, 1975 and 1984*, quoted in Hussein Dimacy, *Patterns of Employment and Incomes in the Tunisian Village* (Tunis, League of Arab States, mimeo (Arabic), 1986). Table 1. Agricultural labour force figures from Table 2.2.

Simultaneously the crude birth rate also fell, but less dramatically: 27 per cent, from 44 to 32 per thousand.[1] The population growth spurted and, at present rates, the population will double in 28 years, as compared with 39 years at the 1956–66 rate. To exacerbate the employment problem, at each period the labour force grew faster than the population, indicating an increase in the female participation rate. Thus between 1975 and 1984 the labour force growth averaged 3.1 per cent, compared with a population growth of 2.5 per cent.

The sectoral figures in Table 2.1 show the nature of the structural transformation undergone by the economy. The year 1975 marks a turning point when the population was almost equally divided between rural and urban areas. Two decades previously, rural areas held twice as much of the population as the urban areas. The turning point in terms of relative share of population in agriculture was reached around 1968. With that, the 'arithmetic of structural transformation' – that is, the effort needed to absorb the incremental population – swung in favour of labour absorption in the non-agricultural sector.[2] For example, when the agricultural population accounted for 75 per cent of the total population, as must have been the case in the late 1940s, a 2 per cent growth in population would have required an 8 per cent growth of employment in the non-agricultural sector to accommodate the growing labour force, whereas with the population evenly divided, as occurred around 1965, a 4 per cent growth would suffice. Thus, the fact worth underlining is that since 1965 or so the non-agricultural sector has borne the brunt of employment creation in Tunisia. Ever since then the proportion of population in the agricultural sector has been declining, reaching only 22 per cent by 1984. In terms of absolute numbers, agricultural population continued to grow until around 1980, but has declined since.

Rural-urban migration accounted for the greater part of the shift in the distribution of the labour force.[3] We note from Table 2.1 that the rate of growth of the urban population amounted to 4.8 per cent over the period 1966–75, compared with only 0.3 per cent for the rural population. Consequently, the urban labour force increased at 6.6 per cent per annum, compared with 2.6 per cent for the rural labour force. The magnitude of the migration implied by these figures is staggering. Had there been no migration, the urban population would have grown by just over one million between 1956 and 1984, whereas in actual fact it increased by 2.4 million. From this one could say that 57 per cent of the urban growth was due to migration and that altogether 1.4 million people left the rural areas for the towns. Put differently, without migration, the rural sector would have had to accommodate over 40 per cent more people than it did.

Rural–urban migration has thus been one of the most important escape routes for the surplus agricultural labour. Rural exodus is a well-entrenched phenomenon in Tunisia. In the past it was associated with various agricultural crises. Thus, the crisis of the early 1930s caused the first wave of rural–urban migration. Later the demise of the co-operative movement in the 1960s led to a second wave of migration, to urban areas as well as to Europe. The present wave of migration, which dates from the 1970s, was associated with rapid mechanization of agricultural production and the development of the tourist industry in the coastal areas. According to the 1984 population census, the inter-governorate migration amounted to 274,860 between 1979 and 1984. Intra-governorate migration amounted to 231,230 during the same period. The total (506,090) represents 7.3 per cent of the population of Tunisia (6,975,450), or 23.7 per cent of the labour force. This points to a high degree of mobility of the Tunisian labour force. This high labour mobility reflects in part the effect of regional differences. The capital and coastal areas developed much faster than the northwest and the south. Rural labour from the disadvantaged areas was pushed to migrate to nearby urban centres, especially during the tourist season. Tunis and the coastal areas were booming, thus creating 'pull' conditions for the surplus labour. In the meantime, as we shall see, increases in agricultural productivity were exerting another type of 'push' on rural labour by making some of it redundant.

Where did the migrants go? More generally, how was the increase in the labour force absorbed? Detailed figures are only available from 1975 onwards but they provide us sufficient material to attempt an answer. These figures are shown in Table 2.2. The labour force increased by 515,000 between 1975 and 1984. According to the traditional sectoral classification, it appears that 80 per cent of this increase was absorbed into employment. To put it differently, employment expanded at the same rate as the labour force, so that the unemployment rate remained practically unchanged at around 16 per cent. There should be no comfort in this because total unemployment increased quite perceptibly. Manufacturing, building and construction absorbed the bulk of the incremental labour force, while employment in agriculture declined in both relative and absolute terms. This is an important phenomenon to which we now turn.

Agricultural labour force

As we know by now, one of the striking features of the employment situation in Tunisia is the rapid decline in agricultural and rural labour force over the last two to three decades. Table 2.1 showed that 1975 represented a turning point at which the population was almost equally divided between rural and urban areas. By 1984 the share of urban areas exceeded that of rural areas. The trend in the labour force was similar. The agricultural labour force registered a decline both in absolute and relative terms (Table 2.2). Between 1975 and 1984 agricultural employment declined by around 7 per cent while its share fell from 37 per cent to 27 per cent.

Table 2.2 Labour force by sectors, 1975 and 1984 (in thousands)

	1975	% of employed	1984	% of employed	Change 1975–84
Labour force	1,622		2,137		515
Employed in	1,367	100.0	1,786	100.0	419
Agriculture	509	37.2	475	26.6	−34
Manufacturing	235	17.2	345	19.3	110
Services	213	15.6	261	14.6	48
Building and construction	128	9.4	237	13.3	109
Administration	n.a.	n.a.	130	7.3	n.a.
Trade	124	9.1	118	6.6	−6
Transport and communications	56	4.1	87	4.9	31
Mining and energy	27	2.0	38	2.1	11
Electricity or gas	11	0.9	n.a.	n.a.	n.a.
Not specified	62	4.5	95	5.3	33
Unemployed[a]	255		351		96
15–60 years	172		245		73
Others	83		106		23
Rate (%)	15.7		16.4		n.a.

Source: INS, *Recensement général de la population et de l'habitat, 1984*, vol. 5. (Tunis, League of Arab States, 1986).
Note: [a] Adjusted to include the unemployed in the age groups 15–17 and 60+.

To understand these shifts, we start by placing agriculture within the context of the domestic labour force – that is, the labour force excluding emigrants. Although Tunisia is still an agricultural country in terms of that sector's employing the largest number of people, agriculture's share in the labour force has declined drastically, to the extent that less than a quarter of the population still finds employment in the agricultural sector. However, as Table 2.3 shows, in the rural areas agriculture remains the most important source of employment, absorbing around 40 per cent of the rural labour force. This being true, perhaps a more remarkable fact is that 60 per cent of rural labour in Tunisia is not employed in agriculture. Some workers remain unemployed – in fact the unemployed constitute the second largest group of labourers in the rural areas – while the rest find work (mainly on a casual and temporary basis) in diverse sectors of the economy, with construction and manufacturing leading. The strength of the construction sector can be taken as proof that the rural areas have shared to some extent in the boom from external sources, especially from remittances. But even the other sectors, while much less developed than in the urban areas, have expanded in the last two decades.

Table 2.3 Structure of the labour force by location, 1984

	Urban	Rural	Total
Labour force			
15–60	1,166,720	970,490	2,137,210
Unemployed	174,930	175,860	350,790
	(15.0%)	(18.1%)	(16.4%)
Employed			
15–60	991,790	794,630	1,786,420
Agriculture	69,900	405,470	475,370
Manufacturing	251,900	92,920	344,820
Services	216,700	44,380	261,080
Building and construction	113,730	123,760	237,490
Administration	100,360	29,150	129,510
Trade	91,500	26,840	118,340
Transport	67,330	19,370	86,700
Mining and energy	29,130	8,900	38,030
Not specified	51,240	43,840	95,080

Source: As for Table 2.2.

The next few tables enable us to probe into some specific aspects of the agricultural labour force. A warning should be noted, however, that these tables are established on a different basis from that of the population census and hence the definition of the labour force differs. For example, in comparing the total agricultural labour force in Table 2.4 (1.12 million) with that in Table 2.2 (0.47 million) we notice an obvious discrepancy. The reason for this is 'double-counting' implicit in the 'agricultural inquiry' method of enumerating labour; for example, temporary labour is quite likely to be counted more than once by its very nature. In addition, temporary labour, because of its intermittent nature, cannot be compared with permanent labour. Thus, our aim in presenting the next series of tables is not to derive some absolute figures but to underline certain important trends.

Table 2.4 shows that wage employment has declined in the agricultural sector and that an increasing number of wage employees work on a temporary basis. Family workers – especially when landholders are included – are much more important than wage employees. A significant proportion of family workers, too, work on a temporary basis, partly because of the low crop intensity and the increasing degree of mechanization in Tunisia's agriculture. In rain-fed areas, machines have replaced permanent wage labourers, but at the same time created bottlenecks at critical periods which can only be overcome by employing labour on a

13

Table 2.4 Agricultural labour force by type, 1976/77 to 1984/85 (in thousands)

| Year | Land holders | Wage labour | | | Family labour | | | Total | No. of days worked |
		Perma-nent	Tempo-rary	Total	Perma-nent	Tempo-rary	Total		
1976/77	357	57	53	110	326	203	529	992	156,554
1977/78	355	61	101	163	344	251	595	1,112	169,733
1978/79	356	51	95	147	323	157	481	983	156,007
1979/80	355	45	102	148	275	298	574	1,077	154,968
1980/81	353	39	77	116	156	287	443	913	126,093
1981/82	355	38	82	120	197	309	506	980	136,095
1982/83	355	37	69	106	201	336	537	998	137,410
1983/84	355	42	69	111	201	336	537	1,003	138,760
1984/85	376	42	75	117	274	360	635	1,128	159,098
Average days/year	160	270	100		200	70		141	

Source: Omar Kaddour, 'Scénarios de projection de la main d'oeuvre agricole', mimeo, January 1986, based on Ministère de l'Agriculture, *Enquête agricole de base*, several issues.

temporary basis. This attests to the fact that mechanization has not been uniform among all agricultural processes. On medium-sized farms, during harvest time at least, the need remains to take on workers to supplement mechanical harvesting and ancillary tasks. Temporary labour by definition points to the existence of underemployment, a condition that is in many ways inherent in all agriculture because of its seasonal character. In Tunisia, the peak at harvest time has been much intensified because of pre-harvest mechanization, and consequently underemployment has increased at non-harvest times. The escape route here has been circular migration, a process whereby rural labour goes to the coastal areas for employment during the tourist season to return to the farms for the harvest, which generally coincides with the slack season in tourism. The situation is different in irrigated agriculture where permanent employment, especially family labour employment, still prevails.

This phenomenon of labour 'casualization' cuts across all farm sizes. Table 2.5 provides the same basic data as in Table 2.4 but broken down by size of landholdings for the one year for which such data are available. The first point to notice is that the majority of landholdings in Tunisia are below five hectares, and these operate almost entirely on the basis of family labour. When they hire labour, it is mostly temporary labour. At the other extreme, the large farms utilize mostly hired labour and usually on a permanent basis. On these farms, as may be surmised from the figures of permanent and temporary labour, mechanization has been pushed even to the stage of harvesting operations. In the two next largest

Table 2.5 Employment by size of landholding, 1979

Size of landholding (hectares)	No. of holders	Per cent	Wage labour				Family labour				Total manpower (= 100%)
			Permanent	Per cent	Temporary	Per cent	Permanent	Per cent	Temporary	Per cent	
0–5	149,000	44.5	2,280	0.7	23,320	6.9	71,550	21.2	89,790	26.7	336,540
5–10	75,800	35.4	3,890	1.8	15,250	7.1	48,470	22.7	70,570	33.0	213,980
10–20	70,600	31.9	2,330	1.1	11,480	5.2	64,730	29.3	72,120	32.5	221,260
20–50	44,100	24.1	6,400	3.5	17,950	10.0	61,250	34.0	51,100	28.4	180,800
50–100	10,500	17.0	5,030	8.3	13,260	21.9	21,670	35.8	10,250	17.0	60,710
100 +	4,400	7.0	25,570	40.5	21,070	33.5	7,720	12.2	4,310	6.8	63,070
Total	355,000	33.0	45,500	4.2	102,330	9.5	275,390	25.6	298,140	27.7	1,076,360

Source: Ministère de l'Agriculture. Enquête agricole de base, 1980.

categories we notice that more temporary labour than permanent labour is employed, attesting to the presence of labour bottlenecks at harvest periods.

Thus, two major conclusions can be reached about shifts in the pattern of labour use in agriculture. First, there was a drop in permanent labour and an increase in temporary labour. This process of casualization applied to both family and hired labour. Second, there was a shift from wage to family labour. Altogether, reliance on temporary family labour increased, as opposed to wage labour in general and permanent wage labour in particular. These trends towards casualization and reliance on family labour cut across all farm sizes. Thus, Tunisia's agriculture has become a part-time activity. No more than a quarter of the total labour force is engaged in agriculture, and even these workers are not fully occupied the whole year round. Farmers themselves are undertaking more non-agricultural activities. In 1985, 43 per cent of landholders had off-farm activities (Table 2.6), most prominent among these being informal activities (denoted by 'others' in the table). These include petty trade and odd-jobbing. This category was followed by construction and trade, with industry and crafts coming at the end of the list. Here again this observation applies to farmers of different sizes of landholdings. As Table 2.7 shows, half the landholders devoted less than two months per year to agriculture and 86 per cent devoted less than six months.

Table 2.6 Non-agricultural activities, 1985

	Numbers	*Per cent*
Total no. of owners	376,000	100
On farm	213,600	57
Off farm	162,740	43
of which		
Agriculture	33,460	21
Industry	9,140	6
Trade	22,700	14
Crafts	3,670	2
Building	33,890	21
Others	59,880	37

Source: Ministère de l'Agriculture, *Enquête agricole de base, 1985*, p. 63.

The question then arises of the interpretation of these shifts in the labour force. Do they denote a transition from an economy based on agriculture to one more diversified? Or do they simply represent 'escape routes' adopted by rural labour to avoid the fate of unemployment in agriculture? The next section attempts an exploration of these issues.

Table 2.7 Distribution of landholders by time devoted to agricultural activities and size of holding, 1980

Size of landholding (hectares)	No. of holders	%	Holders with other activities	%	*% distribution of holders by time devoted to agriculture*			
					0-2 months	*2-4 months*	*4-6 months*	*6+ months*
0-5	149,600	42.1	78,340	55.2	57.3	23.6	9.4	9.7
5-10	75,800	21.4	28,870	20.3	55.5	22.4	12.7	9.4
10-20	70,600	19.9	20,340	14.3	40.0	22.0	16.2	21.8
20-50	44,100	12.4	11,010	7.8	23.2	24.2	19.0	33.6
50-100	10,500	3.0	2,500	1.8	46.8	11.6	13.6	28.0
100+	4,400	1.2	810	0.6	45.7	12.3	3.7	38.3
Total	355,000	100.0	141,870	100.0	51.5	22.9	11.9	13.7

Source: Ministère de l'Agriculture, *Enquête agricole de base, 1980*, December 1980, p. 43.

Traditional escape routes

Three major escape routes were used by the labour force leaving the rural areas: joining the urban labour market, especially the informal sector; emigrating to Europe, and at a later stage, the Libyan Arab Jamahiriya; or joining the ranks of the unemployed in both the rural and urban sectors.

The informal sector

The role of the informal sector in employment creation is the subject of a controversial debate in Tunisia, as indeed in most developing countries. For some,[4] this sector has been very effective in absorbing surplus labour at reasonable levels of productivity, and thus has represented an important source of income for a large proportion of the labour force. For others,[5] the role of the informal sector has to be viewed within the wider perspective of labour market dynamics. Thus, the increased employment in that sector is not viewed as a positive development. It is seen rather as a transitory sector where a labour reserve is created to provide a source of cheap labour supply for the process of modernization. The small-scale industry and craft parts of the informal sector are regarded, according to this view, as a reflection of the dualistic structure of the economy, where the formal sector caters to the demands of the well-off while the informal sector responds to the demands of the poorer sections of the population. Some judicious probing of the data enables us to say something more concrete on this.

17

We start with two estimates of employment distribution by the type of organization: private, public, modern and informal (Table 2.8). A common problem with such estimates, as should be evident from the two quoted here, is that they ultimately depend on the definition adopted of the 'informal sector'. In the two estimates shown, the informal sector is divided into two parts: localized and non-localized. The localized sector consists of enterprises employing fewer than 10 persons (and quite likely more than five) while the non-localized sector is obtained as a residual, presumably to mean enterprises employing one to five persons. In that case, the modern private sector should consist of enterprises employing more than 10 persons. In practice, as can be ascertained from the estimating procedures shown by Charmes,[6] the methodology of

Table 2.8 Labour force breakdown by major forms, 1975, 1980 and 1984 ('000 and percentages)

	1975	1980	1984	1975 to 1984	
				Change ('000)	Percentage change
A. Charmes estimates					
Agriculture	509	539	475	−34	−6.7
Administration	145	191	271	126	86.9
Public enterprises	127	177	210	83	65.3
Modern private sector	133	186	238	105	78.9
Informal sector	329	374	430	101	30.7
Localized	155	178	205	50	32.2
Non-localized	175	196	225	50	28.6
Seasonal	61	51	68	7	11.5
Not specified	62	59	95	33	53.2
Total	1,367	1,577	1,786	419	30.7
B. Abdel-Jelil estimates					
Agriculture	509	558	475	−34	−6.7
Administration	145	191	239	94	64.8
Public enterprises	114	155	179	65	57.0
Modern private sector	164	198	230	66	40.2
Informal sector	369	423	568	199	53.9
Localized	141	188	282	141	100.0
Non-localized	228	235	286	58	25.4
Not specified	66	59	94	28	42.4
Total	1,367	1,577	1,786	419	30.7

Sources: A: 1975 and 1980 from J. Charmes, 'Deux estimations de l'emploi dans le secteur non structuré en Tunisie', in *Séminaire sur les statistiques de l'emploi et du secteur non structuré* (Paris, 1985), vol. 2, Table 1, p. 442.; 1984 from Charmes, 'Principales tendances de l'emploi et du chomage en Tunisie, 1956–1984', UNDP/ILO report, in press).
B: Bedoui Abdel-Jelil, *L'emploi non-agricole et urbain en Tunisie* (Tunis, League of Arab States, mimeo, 1986), Table 18.

deriving such estimates is very inexact and a great deal of data shuffling has to be done to arrive at such categories. Thus, differences are to be expected. The two authorities quoted here differ by a great deal, particularly with respect to the informal and modern sectors. Part of this discrepancy arises because Charmes (estimate A) has an extra category of workers – seasonal workers in the tourist and building industries – whereas Abdel-Jalil (estimate B) includes all of these in the informal sector. Both end up with about the same number of workers in the modern private sector, but as Charmes starts with a lower figure he gets a perceptible growth. On the other hand, for Abdel-Jalil the modern sector's share remains constant. According to Charmes it is the informal sector that maintains its share.

Given such differences in estimates, a question arises concerning the exact role of the informal sector in the absorption of the labour force. According to Charmes, the informal sector accounted for around one-fifth of the employment opportunities created in the non-agricultural sectors during the decade 1975–84, whereas it created twice as many opportunities according to Abdel-Jelil. Even on the lower estimate the sector employed around one-third of the non-agricultural labour force in 1984. This points to the importance of the sector in the process of structural transformation as a substantial proportion of the urban labour force was absorbed in a growing informal sector.[7] Tunisia, in this respect, presents no exception to the patterns observed in other developing countries, especially in Asia and Latin America.

If we add this information to the type of employment created in the informal sector, we can see that there were many positive aspects to the structural transformation process in Tunisia. Here it is important to recognize that the informal sector in Tunisia maintained its relative share of employment during the period considered. Second we note that this sector (as in the other Maghreb countries) is the norm rather than an aberration. As Charmes puts it:

> Dans un pays aussi anciennement urbanisé que la Tunisie, l'artisanat de production et de service, et le petit commerce, ont toujours représenté une fraction importante de l'emploi non agricole, depuis des dates très anciennes. [In a country of such a long history of urbanization as Tunisia, small-scale production and service crafts as well as petty trading have always constituted an important share of non-farm employment.][8]

Thus, in Tunisia it is the informal sector on which the modern sector has been superimposed and not vice versa as, for example, is the case in most sub-Saharan African countries.[9] It is by and large a vibrant sector, comprising labour of many different skills and in some cases providing a training ground for skill formation, as may be seen by looking at the breakdown of the non-agricultural labour force in 1980 (Table 2.9).

Table 2.9 Informal sector employment by localization compared with public and private modern sector employment, 1980 ('000)

	Informal sector			Public sector	Private modern sector	Total
	Localized	Non-localized	Total			
Mines and energy	n.a.	n.a.	8.2	29.2	3.3	46.7
Manufacturing industry	n.a.	n.a.	151.5	62.3	91.9	299.9
Textile, clothing, leather	15.1	91.4	106.5	19.9	44.9	n.a.
Wood	16.8	5.5	22.3	5.7	9.0	n.a.
Building, public works	1.3	40.7	42.0	9.1	37.1	158.1
Commerce	79.6	10.8	90.4	9.5	15.0	106.3
Services	48.4	33.5	81.9	66.9	38.6	193.8
(Administration)	n.a.	n.a.	n.a.	n.a.	n.a.	(190.5)
Total	178.4	195.6	374.0	n.a.	n.a.	1,037.5

Sources: Derived from J. Charmes, 'Emploi et revenus dans le secteur non structuré des pays du Maghreb et du Machrek', conference on the Informal Sector in the Middle East and North Africa, Tutzing, Federal Republic of Germany, 28–31 July 1986, Table 1, p. 10; Charmes, 'Secteur non structuré: Politique economique et structuration sociale en Tunisie, 1970–1985', Table 2, p. 12; and Charmes, 'Deux estimations de l'emploi dans le secteur non structuré en Tunisie', in *Seminaire sur les statistiques de l'emploi et du secteur non structuré*, vol. 2 (Paris, 1985), Table 4, p. 447.
Notes: Row figures do not total up as the category 'Administration' has been separated out in the source. Column totals do not total up due to the inclusion of only selected items. N.a. indicates data not available in the sources.

Several important facts about the informal sector emerge from Table 2.9. First, unlike most other developing countries, a majority of those in the informal sector in Tunisia were not working in the commercial sector, but in the manufacturing sector. In fact, of the total employment in the manufacturing sector, one-half was located within the informal sector. Second, as might be expected, with firms employing fewer than 10 persons being the rule rather than an exception in the commercial sector, the informal sector almost completely dominated that sector. However – and this is a significant finding – only 10,800 of the persons counted here were put in the non-localized category. These would comprise what in most other countries is thought of as the informal sector: vendors, hawkers, shoe-shine boys, and so on. In fact, the non-localized category dominated in the manufacturing sector, particularly in the clothing industry. This might conjure up visions of foot-loose operators of small machines circulating from place to place offering their wares, but the story is rather different. The 91,400 persons – the highest – entered as non-localized against the clothing industry were all entered in the background table on which these estimates are based as *travail à domicile* – that is, home-based work is counted in the non-localized category.[10] From this we may infer that the division between localized and non-localized labour is effectively in terms of workshop versus home operation, and not fixed-address versus mobile operation – and certainly not low productivity versus high productivity. In Tunisia, as in many developing countries, this type of 'putting-out system' has emerged as an important survival strategy for the labour force. Similarly, we find that 40,700 non-localized workers were entered against *bâtiments, travaux publics*; these were truly non-localized in the sense that they were itinerant workers who went where employment opportunities took them.

The importance of the informal sector in Tunisia can also be gauged from its contribution to incomes. Table 2.10 throws light on this in terms of the ratio of value added per head in the informal sector as compared with the modern sector. In the dominant informal sector activities (textiles, commerce and woodwork), value added per head (or productivity) was comparable to that in the modern sector (textiles and wood), or not too far behind (commerce). In the case of textiles, we may recall that most of the industry was non-localized; this once again confirms our point that there is no necessary correlation between an activity's being non-localized and its having low productivity. The impression from this Table is that the informal sector in Tunisia is a fairly productive one in comparison with the modern sector and thus its growth should not be regarded *ipso facto* as absorption into a low-productivity trap.

Table 2.10 Ratio of value added per capita in informal sector compared with modern sector, 1981

Agriculture and food	0.38	Commerce	0.62
Construction materials	0.52	Services	0.63
Mechanical, metallic	0.49	Total	0.70
Textile	0.76		
Wood	0.85		
All five above	0.59		

Source: J. Charmes. 'Emploi et revenus dans le secteur non structuré de pays du Maghreb et du Machrek', conference on the Informal Sector in the Middle East and North Africa, Tutzing, Federal Republic of Germany. 28–31 July 1986, Table 7, p. 23.
Note: The table should be read as follows: value added per capita in 'agriculture and food' in the informal sector was 38 per cent of the corresponding figure in the formal sector.

But does the growth of the informal sector not imply increasing underemployment? Has not the informal sector been absorbing more labour than it can do productively? There are two implicit assumptions behind this belief: (1) that most of the labour force in the informal sector is composed of family members, so that its expansion is simply to accommodate new relatives from the countryside rather than to hire labour to meet growing demand; and/or (2) that most workers in the informal sector are poorly paid, and worse, their wages are continually pushed down to counteract falling productivity. The fact is that in Tunisia the informal sector does not consist mostly of family members. It has been estimated that in 1980, 31 per cent of the labour force in the informal sector consisted of salaried workers and 25 per cent of paid apprentices. The rest (44 per cent) were owners and family helpers.[11] If we assume that these were equally divided, we get at the most 25 per cent dependent family workers in the informal sector. The fact that well over half of the labour force was paid a cash wage implies that they were hired because they had something positive to contribute, and not to keep them off the streets – unless of course it can be shown that wages were low and falling. Data are not available on the latter, but on the former they contradict strongly any notion that workers in the informal sector were being exploited. Thus, taking the SMIG – the minimum wage in towns – as a standard, it has been estimated[12] that in 1981 in manufacturing industry (excluding food industry)[13] qualified workers earned 1.2 times the SMIG, whereas semi-skilled and unskilled workers earned 0.8 times the SMIG. For the two categories of workers combined, the average wage was 7 per cent higher than the SMIG. In textiles and wood, which are the largest industries, the average wage in the informal sector was respectively 4 and 7 per cent higher than the SMIG. Similarly, entrepreneurs' earnings in the manufacturing industry (again excluding food) were 5.6 times the SMIG, in commerce 4.4 times, and in all services combined 8.9 times.[14] One certainly does not get a notion here of masses of impoverished workers. We shall see a further confirmation of this in Chapter 4.

Having interpreted the available figures, some qualifications should now be noted. First, the informal sector in Tunisia (and other Maghreb countries) is structurally different from that in other developing countries, especially in sub-Saharan Africa. In the Maghreb, the informal sector is the traditional sector, whereas in sub-Saharan African countries it is a new phenomenon. In the Maghreb, it is the informal sector from which the modern sector carves out its market, whereas in the sub-Saharan African countries it is the informal sector that is constantly trying to encroach upon the modern sector, but mostly succeeds at the petty end of trading activities. There are also huge differences in skill levels in the informal sector in the two types of African countries.

This point being made, we should note that the Tunisian informal sector was operating under conditions of a growing market in the last decade. Thus, its task of absorbing incoming migrants was made that much easier. Demand was growing and labour could be deployed to meet this demand. Now that the demand conditions have changed, will the informal sector be able to continue to absorb labour productively, as its position as the dominant sector requires it to do? More broadly, does the informal sector have the potential to evolve into a high-productivity sector? Does it show promise of producing entrepreneurs who can adopt new forms of organization and new technologies, or is it merely replicating itself – sons taking over the family business from their fathers and running it along traditional lines? The point is, while 'traditional' does not mean 'low productivity' and 'modern' does not imply 'high productivity', there is no doubt that in all countries at all stages of development new forms of production have to evolve to spearhead the process of development. Is the informal sector in Tunisia capable of this? On past experience the verdict must be negative. Take the example of the clothing industry. Both forms of organization exist in this industry but they remain separate: the traditional sector makes traditional clothes (*jelabiya*, or robes) for the local market while the modern sector makes jeans for the export market. The traditional industry has remained sewing-machine-bound while the modern industry operates as an enclave, making few efforts to mechanize the traditional clothing industry. Thus doubts must remain about the ability of the informal sector to spearhead development, given its demonstrated lack of dynamism in the past.

It is against this background that the controversy about the informal sector should be viewed. Whatever point of view we may wish to hold, the important issue is to what extent the informal sector can be relied upon to generate productive employment in the future. The answer that emerges from the foregoing analysis is that, looked at in a dynamic framework, the informal sector cannot be expected to continue to play the same role in labour absorption. Its limits as an escape route for new entrants to labour markets have quite likely been passed.

Emigration

Emigration provided another escape route for the displaced agricultural labour force. The recent wave of emigration dates back to the early 1960s when large numbers of the Tunisian labour force sought employment in the labour markets of Europe, especially France. By the late 1970s, net return migration from Europe began to increase and the Libyan Arab Jamahiriya had emerged as the main destination for Tunisian labour. Table 2.11 shows the estimates of official emigration from Tunisia. These figures have to be treated with caution since they represent officially recorded migration and therefore probably underestimate the actual flow of emigration. Estimates of the number of emigrant workers vary from 150,000 to 300,000 (see Appendix A). According to a study by the Ministry of Social Affairs, the total number of Tunisian emigrants in 1985 was thought to be 280,000, or some 11.6 per cent of the total

Table 2.11 Net emigration and remittances, 1964–86

Year	Net emigration	Official emigration	Remittances (million dinars)
1964	10,245	n.a.	2.6
1965	11,411	n.a.	3.8
1966	12,637	n.a.	4.6
1967	14,476	n.a.	5.9
1968	17,731	n.a.	7.7
1969	27,456	7,840	11.4
1970	21,906	13,808	15.2
1971	32,481	14,658	22.7
1972	24,559	16,319	29.6
1973	12,768	18,947	41.2
1974	−2,352	8,620	51.7
1975	2,135	4,761	58.2
1976	−19,200	2,381	61.4
1977	34,800	2,817	72.2
1978	13,000	28,908	91.8
1979	n.a.	13,550	115.5
1980	n.a.	3,809	129.2
1981	n.a.	7,693	178.3
1982	n.a.	8,654	219.6
1983	n.a.	4,379	235.0
1984	n.a.	1,181	246.0
1985	n.a.	n.a.	210.0
1986	n.a.	n.a.	190.0

Sources: World Bank, *Tunisia: Social Aspects of Development* (Washington, DC, June 1980), p. 105, for 1964–78; for 1979–83, Ministère des Affaires Sociales, Office de la Promotion de l'Emploi et des Travailleurs Tunisiens à l'Etranger, *Statistiques du marché du travail et des placements à l'étranger*, (Tunis, September 1985); 1984–86 from data provided by Ministry of Planning.
Note: n.a. = not available.

(external and internal) labour force. Estimates made in the Sixth Development Plan show that between 1962 and 1981, emigration absorbed almost 29 per cent of the incremental labour force (Table 2.12). In the first decade, the rate was twice as high as in the second decade. Projections for the period 1982–91 were that emigration would further decline, and only under 10 per cent of the incremental population would find an outlet through this route.

Apart from providing an escape route, the importance of emigration lies in the remittances sent back to Tunisia. Table 2.11 contained estimates on this. Combining these with the data in Table 1.1. we can see that in 1984, despite the growing importance of other external funds, remittances still provided 21 per cent of externally generated resources and comprised just under 4 per cent of the GDP. Although precise data are

Table 2.12 Contribution of emigration to employment, 1962–81 and
projections for 1982–91 ('000)

		1962–71	*1972–81*	*1982–91* *(projections)*
1	Net addition to the labour force	357	469	664
2	Emigration (of 18–59 years)	140	97	50
3	2 ÷ 1 (%)	39	21	8

Source: République Tunisienne. *VIème Plan de développement économique et social
(1982–1986)*. vol. 1, p. 150.

not available, it is generally reckoned that a large part of the remittances
go to rural areas because of the rural origin of most emigrants. Thus,
in a study of migration undertaken in 1976, it was estimated that remittances from migrant workers represented 20 per cent of the household
consumption in the rural areas of Tunisia.[15] The recent trend of return
migration from Europe and expulsions from the Libyan Arab Jamahiria
(in September 1985, 32,000 Tunisian workers were expelled) point to
grave effects on the Tunisian rural economy in terms of household income
as well as labour absorption. The disadvantaged areas of the north-west
and the south are particularly affected as they contribute most of the
migrant labour. These trends may be mitigated by the recent reopening
of the Libyan labour market for Tunisians, and the beginnings of emigration towards the service sector of other Arab oil-producing countries.

Unemployment

Despite the escape routes provided by emigration and employment
expansion in the informal sector, unemployment increased in absolute
terms in the last decade. Even the rate reached in 1984 – 16.4 per cent
(see Table 2.2) – although remaining practically unchanged over a
decade, was high by developing country standards.[16] Moreover, the
figures as obtained from the census mask in some respects the actual
unemployment problem (see Appendix B). First, the questionnaire used
for the 1984 population census asked a supplementary question to 'Are
you working?': 'Do you own land?' If the answer to the second question was 'yes' and to the first 'no', the respondent was classified as
employed, presumably on the grounds that he could always return to
his land to escape urban unemployment. That is partially true, but the
fact is that in Tunisia most urban workers could say they owned land
in the rural areas, referring to family-owned land, but not all of them
would necessarily return to it to escape unemployment due to the

25

extremely low productivity of this land. Thus, the supplementary land question minimizes the true extent of unemployment by some 90,000 people.

Second, in the female labour force only 20,000 or so were counted among the unemployed, compared with ten times as many among males. The reason for this is that most women who said they were unemployed were entered as housewives rather than as unemployed, although they may have had some occupation such as home-based handicrafts. Here again, by minimizing the total number of females in the labour force, the census effectively underestimated the unemployment rate.

Third, the question put to the population was 'Are you working today?' If the answer was 'yes' the respondent was entered as employed. The respondent might have just ended unemployment of long duration but that fact would be masked. In the opposite case, a person might have recently joined the ranks of the unemployed, or he may have simply left his job in search of another; he would be counted as unemployed. Thus, unemployment figures tell us only how many people in the labour force were counted as not working on the day of the census. From this one is at a loss to judge the severity of the unemployment problem. For example, we would like to know the percentage of the unemployed according to the duration of their unemployment. From background data in the population survey of 1980, we find that over 60 per cent of the unemployed were out of work for more than six months.[17] In contrast, those who were unemployed for less than one month were just over 10 per cent. Thus, we may surmise that the unemployment problem in Tunisia is not transitory and affects around 10 per cent of the population on a long-term basis. Later censuses indicate that the severity of the employment problem measured thus has increased. For example, in the 1983 unemployment survey, 68 per cent of the unemployed were unemployed for six months or longer.[18]

The majority of the unemployed come from poverty-stricken areas of Tunisia (Table 2.13), the highest rate of unemployment (24 per cent) being recorded in Jendouba, followed by Le Kef (19 per cent), Gafsa (18 per cent) and Kassérine (17 per cent), all of which are among the poorest districts of Tunisia. Moreover, as was to be expected, unemployment was highest among the uneducated. In 1984, 35 per cent of the unemployed had no education, while 46 per cent had access to only primary education.[19] This is not to minimize the incidence of unemployment among the educated; data show that the rate of unemployment among the latter category amounted to 20 per cent. Finally, and perhaps most importantly, unemployment is primarily a youth phenomenon. In 1984, 71 per cent of the unemployed belonged to the age group 15–24.[20] The same sort of figure was found in the 1975 census.[21] Thus,

Table 2.13 Regional characteristics of active population and unemployment,
1975 and 1984

	Active population ('000)		Unemployment as % of local labour force	
	1975	*1984*	*1975*	*1984*
Tunis	294,360	276,950	12.9	12.8
Zaghouan	56,920	37,980	15.3	11.6
Bizerta	97,830	125,800	20.0	16.0
Béja	79,670	89,970	18.5	15.2
Jendouba	81,580	114,200	32.8	23.7
Le Kef	65,500	75,030	20.6	19.1
Kassérine	79,940	82,210	15.5	16.9
Médenine	79,710	72,770	9.7	8.2
Gabès	68,860	66,820	16.0	15.7
Gafsa	61,670	60,890	17.1	18.4
Sidi Bouzid	50,420	77,270	16.2	13.8
Kairouan	97,160	122,610	21.3	10.9
Sousse	74,420	100,010	12.4	11.2
Sfax	127,480	173,800	11.5	9.8
Nabeul	116,390	152,090	9.8	8.0
Monastir	71,620	88,230	12.6	11.6
Mahdia	65,760	83,630	8.7	10.3
Siliana	53,430	65,630	25.7	16.5

Sources: INS. *Recensement général de la population, 1975,* pp. 180–97; and INS. *Recensement général de la population et de l'habitat, 1984,* pp. 111 and 193.

we can say that new entrants to the labour market found it difficult to
obtain jobs – something that is bound to become more and more true
in the future.

Conclusion

The trends described in this chapter point to important shifts in labour
use and labour contracts in Tunisia over the last three decades. The share
of agriculture in the labour force declined both in absolute and relative
terms, but even for those counted in the agricultural sector, agriculture
became a part-time activity. Do these shifts imply the inability of the
agricultural sector to provide a livelihood for the majority of the population? The answer depends on an understanding of the factors that have
influenced agricultural productivity. Although a full answer must await
the discussion in the next chapter, four such factors may be identified:
(1) trends in agricultural productivity; (2) changes in the land tenure
system; (3) shifts in cropping patterns; and (4) government investment
policy. The point to be emphasized here is that over the years, increasing mechanization and a decline in traditional livestock farming combined

to reduce the demand for labour. The former resulted primarily from a massive investment programme by the government in agriculture and centred on a highly subsidized programme of mechanization and credit. This meant that fewer labourers were required to feed the growing population. One can speak of an agriculture in transition along the lines described by Kuznets. However, the role of other 'push and pull' factors should not be minimized. Studies on internal and external migration point to the fact that the government programmes favoured the large farmers and therefore productivity in the smallholder traditional sector lagged behind. It is from that sector that many young people were pushed into migration. The concentration on the development of the coastal areas for tourism and on industrial activities provided the outward manifestations of the pull to attract labour from rural to urban areas, but in neither case was the expansion in employment opportunities sufficient to absorb the migrating labour force. In the end, despite the safety valve provided by emigration, unemployment increased. The situation would have been even graver had not the traditional sector absorbed a great part of the growing urban labour force. Under the impetus of the all-round increase in incomes in the last decade, productivity levels in that sector remained comparable to those in the modern sector, but it would be too much to expect this situation to endure over the long term, given the recent contraction in the market. Thus, all the signs are that the employment situation will worsen in the years to come, unless steps are taken to correct some of the basic trends in the economy. What these are we shall discuss in the final chapter, after we have completed our review of the agricultural sector and of the equity situation.

Chapter three

The agrarian economy: Evolution and prospects

At the time of independence, agriculture was the major sector in the Tunisian economy. In 1960 it accounted for 24 per cent of GDP, 56 per cent of employment and 51 per cent of export revenues. Since then, however, its importance has been declining rapidly so that by 1984, it contributed only around 14 per cent of the GDP, 27 per cent of employment and less than 10 per cent of export revenues. These tendencies are, of course, normal correlates of the growth process in any developing economy. In Tunisia they were even more to be expected because of the developments in the oil sector, in remittances and tourism. The disquieting fact is that agriculture seems to have gone into something of an absolute decline and the process of structural change has been associated with an increasing inability of the agricultural sector to supply the economy with either enough food or a significant investible surplus. External trade in agricultural products, which produced a healthy surplus in the 1960s, has been generating a growing deficit since the early 1970s and agricultural production has become increasingly dependent on state investment and subsidies. These facts, of course, need to be viewed within the context of the overall pattern of development in the country. As shown in Chapter 5, many different factors have contributed to the food imbalances – not the least of which was a shift in demand from traditional to non-traditional cereals. Also relevant is the central theme of Chapter 2 – the combination of factors which led to a decline in the agricultural labour force in absolute terms during a part of this period. All this, however, does not mask the fact that agriculture has performed erratically since independence and very poorly since 1976. Moreover, it should be recognized that the tendencies discussed in Chapters 2 and 5 were themselves in part the consequences of the poor performance of agriculture and that all this occurred even though there was no discernible anti-agriculture bias in state policy in terms of resource allocation.

Evolution of agrarian policy

We start with a brief account of the evolution of state policy since independence. Four subperiods can be identified from this standpoint. Between 1956 and 1960 government policy was primarily concerned with replacing the traditional collective system of landholding (under which peasants had only usufructuary rights to land) with a system of private property rights in land. This policy effectively transformed the occupants into owners. The state also undertook a significant amount of investment in capital construction in agriculture (afforestation, land improvement, irrigation development, etc.) during this period.

Two important shifts in state policy occurred in 1961. First, a process of establishment of state farms through the acquisition of the land hitherto occupied by colonial settlers was set in motion. Second, a programme of co-operativization of agriculture was launched. The idea was to transform gradually the state farms into co-operatives by incorporating the surrounding peasant farms. The development of co-operatives, it was hoped, would create conditions for efficient use of labour, technological transformation and planned development of agricultural production. State investment was significantly stepped up, and in allocating investment funds (including credit) priority was increasingly accorded to the co-operatives.

On the face of it, the programme of co-operativization was sound enough; its implementation, however, proved difficult. The programme depended on foreign aid for finance (more than a third of the state investment during the period 1962–71, for instance, was supposed to be funded by aid from international organizations). Some classic mistakes were made. A large part of the investment went into projects with long gestation lags and little attention was paid to short-run production growth. The production performance of agriculture suffered in consequence. The members of the co-operatives were dissatisfied and the large landowners were, not surprisingly, hostile to the programme. The final blow came from the World Bank which refused to provide aid for the proposed plan for 1969–72.

By the end of 1969, the co-operativization programme was effectively abandoned and during the next five years or so, state policy towards agriculture was gradually reoriented. The land belonging to the co-operatives and to the state was privatized, direct state investment was gradually reduced and replaced by cheap credit and subsidies to private farmers, and the allocation of investment was increasingly biased towards the requirements of short-run growth. These policies had some initial success, with agricultural production achieving the best performance of any period. In part, though, this also reflected the coming to fruition of past investments.

Also occurring during this period were the oil boom, the emigration of labour out of the country and the consequent rapid growth of remittances. These factors produced three kinds of influences on state policy towards agriculture. First, the state no longer needed to rely on the agricultural sector for either food supply or supply of investible resources. Second, the state could distribute subsidies to agricultural producers in the hope of promoting capitalist entrepreneurship in agriculture. Third, migration of labour out of agriculture (both to other countries and to the rapidly expanding industrial and service sectors within the country) reduced the pressure on the state to address the land problem.

The rapid growth of incomes caused important shifts in demand towards 'luxury' foods. Under the expansive mood created by the foreign exchange boom the state put its trust in the large farmers to feed the growing and ever more affluent urban population. The result was a predictable decline in traditional farming (particularly in the livestock sector) on the one hand, and, on the other, a growing dependence of agriculture on imported inputs, subsidies and cheap credit. Superficially, capitalist farming developed, but without its basic characteristic – generation of self-sustained process of investment and technical change. The overall production performance of agriculture during the period was also poor.

The fall in oil revenues and remittances exposed the weaknesses of the structures created. With the decline in the state's capacity to import inputs and to subsidize agricultural production there has been a general decline in investible resources and a halt in income growth. In the meantime, urban unemployment has been increasing, exacerbated by the return of migrant workers from abroad. In the foreseeable future, the withdrawal of labour from agriculture is unlikely to be feasible. In fact agriculture will have to absorb much of the incremental labour force and play a much more dynamic role in the development process than it did during the boom years. It will have to do all this with a much lower level of state support. This is the context which must define the nature of state policy towards agriculture at the present juncture.

Trends in agricultural production

Given the available data base, trends in agricultural production can be studied in some detail only for the period 1962–84. For purposes of analysis this period has been subdivided into three subperiods: 1962/64–1969/71, 1969/71–1976/78 and 1976/78–1982/84, which coincide approximately with different phases of state policy. During the first subperiod, a co-operativization strategy was pursued; the second subperiod constituted a period of transition from a co-operative-based strategy to a private-farmer oriented strategy; and during the third subperiod a strategy of state-sponsored development of capitalism in agriculture was pursued. The relevant data are presented in Table 3.1. From the point of view of growth of agricultural output, the co-operativization strategy was obviously a failure. In fact, aggregate agricultural output must have declined during the co-operative period, 1962/64–1969/71. This could be attributed partly to the disruption generated during the process of implementation of the co-operativization programme, and partly to the associated policies relating to state investment. As we shall see below, state investment in agriculture was high during this period but primarily financed irrigation projects with long gestation periods.

Table 3.1 Average annual rates of growth of agricultural output, 1962/64 to 1982/84, selected years

	1962/64– 1969/71	*1969/71– 1976/78*	*1976/78– 1982/84*
Wheat	−1.4	6.8	0.9
Barley	−3.5	4.9	9.9
Vegetables	3.6	6.6	3.4
Fruit	−2.4	11.6	−1.3
Industrial crops[a]	−4.0	15.2	1.8
Pulses[b]	2.7	12.0	2.4
Meat[c]	n.a.	7.2[d]	−2.0
Chicken	n.a.	14.0	9.5
Fish	5.0	8.0	4.2
Total agricultural output (constant prices)	n.a.	5.1[d]	1.8

Sources: Estimated from data provided in Ministère de l'Agriculture, *Annuaire des statistiques agricoles, 1984* (Tunis, 1985); and IBRD, *Tunisia: Agricultural Sector Survey* (Washington, DC, 1982).
Notes: [a] Sugar-beet and tobacco
[b] Broadbeans and chickpeas
[c] Beef, lamb and goatmeat
[d] Refers to the period 1971–1976/78

In contrast, agricultural growth was quite impressive during the period 1969/71–1976/78, part of this signifying the pay-off from investment in the previous period. Thus the irrigated area expanded quite rapidly, the use of chemical fertilizers also increased, and high-yielding seeds were introduced.

Two observations can be made for the period 1976/78–1982/84. First, there was a stagnation in agricultural output. Second, there was a change in its composition: in terms of crops, there was a clear shift away from the staple cereal (wheat) and fruit to barley (which is used as animal feed), pulses and vegetables. Poultry farming recorded impressive growth. Rather surprisingly, in view of the other trends and all-round growth in incomes, meat production registered a decline, the expansion of modern livestock farming failing to counteract the decline of traditional livestock farming (as we shall see below). Even so, one could say that there was a shift away from traditional crops to livestock farming and high-value crops. (Traditional livestock farming relies on natural pastures and hence is not linked to the cropping pattern, while modern livestock farming requires the production of barley and forage crops.)

Table 3.2 presents data on the evolution of the cropping pattern. The area under vegetables and pulses increased steadily while that under industrial crops stagnated throughout the period under consideration. The area devoted to forage crops also grew, particularly between 1972 and 1978. While during the co-operativization phase (1962–72) the area under wheat expanded at the expense of that under barley, during 1978–84 the trends were reversed. The area under fruit fluctuated wildly, which is rather surprising since trees do not grow or die so quickly.

Table 3.2 Cropping pattern, 1962 to 1982–84, selected years ('000 ha)

	1962	1972	1976–78	1982–84
Wheat	1,190	1,227	1,322	899
Barley	585	419	461	578
Vegetables	31	74	95	116
Fruits	1,310	1,896	1,492	1,870
Industrial crops	8	6	7	7
Pulses	80	84	93	114
Forage	35	88	259	292
Total	3,239	3,796	3,629	3,876

Sources: Ministère de l'Agriculture, *Annuaire des statistiques agricoles, 1984* (Tunis, 1985), and *Enquête agricole de base* various years; M. Boudhiaf, 'Desertification de la campagne Tunisienne', (Tunis, 1986), mimeo.

The overall trends in the cropping pattern can be summed up as follows. There was a long-term rising trend in the area under vegetables and pulses, reflecting the combined effect of increasing profitability and expansion of the irrigated area (vegetables are cultivated primarily on irrigated land; see Table 3.3). The other important trends are the very rapid growth of area under forage crops since 1972 and the shift from wheat to barley since 1978. The evolution of the cropping pattern thus indicates that there was a shift to modern livestock farming; this was in conformity with the growing importance of livestock products in the average diet and with the state policy of encouraging modern livestock and poultry farming.

Table 3.3 Allocation of irrigated area, 1983–84

Crop	Irrigated area allocated		Irrigated area as % of area cultivated under the crop
	'000 ha	Per cent	
Wheat and barley	15	8.2	1.0
Forage crops	18	9.8	6.0
Pulses and industrial crops	13	7.0	7.4
Vegetables	95	51.6	73.6
Fruits	43	23.4	2.3
Total	184	100.0	4.7

Sources: Ministère de l'Agriculture, *Enquête périmètres irrigués, 1985* (Tunis, 1986), and *Enquête agricole de base, 1984.*

Data presented in Table 3.4 bear out the failure in the livestock sector. Neither the policy of co-operativization nor the policy of encouraging modern livestock farming was helpful: livestock population declined quite sharply between 1956 and 1970 and again between 1976 and 1984. In between there was an increase. It is known that attempts to organize the peasants into co-operatives led to a large-scale slaughter of animals. The explanation of the more recent trends is somewhat problematic. The sharpest decline during the last period occurred in the cattle population. Two contradictory forces seem to have been at work simultaneously: as the cattle population in modern livestock farms increased (because of the policy of supplying highly subsidized feed concentrates), that in the traditional sector declined (because of a decline in pasture land and the growing control of forage crops by the larger landholders). The latter trend obviously dominated over the former; we shall revert to this point at a later stage. We should note that the poultry population increased at a rapid rate throughout the period 1970–84.

Table 3.4 Livestock and poultry population, 1956 to 1982–84, selected years ('000 head)

	1956[a]	1970[a]	1976–78	1982–84
Cattle	736	560	860	594
Sheep	4,335	4,161	5,727	5,285
Goats	1,800	628	1,049	996
Total	6,871	5,349	7,636	6,875
Poultry	n.a.	4,000	12,700	26,900

Sources: Ministère de l'Agriculture, *Annuaire des statistiques agricoles, 1984* (Tunis, 1985), and *Budget économique, 1986* (Tunis, 1986); M. Boudhiaf, 'Desertification de la campagne Tunisienne' (Tunis, 1986); IBRD, *Tunisia: Agricultural Sector Survey* (Washington, DC, 1982).
Note: [a]Livestock numbers are authors' estimates based on data on the female population.

Technological change

Technological change in Tunisia's agriculture has principally involved growth of fertilizer use, tractorization, irrigation and high-yielding seeds, particularly wheat. In other words, one could say that most of the important vehicles of technological change – both yield-increasing as well as acreage-expanding – have been deployed. The relative emphasis has varied at different times. Some relevant data are presented in Table 3.5.

Table 3.5 Indicators of technological change, 1960–62 to 1982–84

	1960–62	1969–71	1976–78	1982–84
Fertilizers used (kg/ha)	11	23	42	60
Tractors per 1000 ha	n.a.	4.7	5.3	6.4
Irrigable area/				
cultivated area	n.a.	3.2	4.3	4.7
Total irrigable area				
('000 ha)	n.a.	120	156	184
HYV-seeds of wheat used				
(kg/ha)	n.a.	n.a.	14.3	32.7

Sources: Data for 1960–62 and 1969–71 from IBRD, *Tunisia: Agricultural sector survey* (Washington, DC, 1982); remainder from Ministère de l'Agriculture, *Enquête périmètres irrigués 1985* and *Enquête agricole de base*, 1980. Data on HYV-seeds from Ministère de l'Agriculture, *Annuaire des statistiques agricoles, 1984* (Tunis, 1985).

Unfortunately, because of lack of data for the initial period, we cannot say much about the nature and pace of technological change during the co-operativization phase (1960/62–1969/71). The only clear trend is that fertilizer use increased rapidly during this period. There is also some indication that tractorization, too, progressed rapidly. On rough estimates, the number of tractors rose from around 10,000 in 1961/62 to more than 17,000 by 1971. As for irrigation development, the only available information is that large investments were made for the purpose during the period (see the section on investment which follows). High-yielding seeds were introduced only in the 1970s.

The periods 1969/71–1976/78 and 1976/78–1982/84 did not differ significantly in terms of the pace and pattern of technological change. The new element was the introduction of high-yielding seeds of wheat in the early 1970s and their rapid spread thereafter. Progress was also made in the use of fertilizers, in tractorization and in irrigation development.

Not much attention was paid to the livestock sector during the early period, 1960/62–1969/71. As noted before, co-operativization had provoked a large-scale slaughter of animals with a consequent decline in the livestock population. From the mid-1970s onwards, the state sought to develop modern livestock farming by encouraging the production of forage crops, by subsidizing the purchase of imported breeds of animals and by subsidizing the production and import of feed concentrates. Unfortunately, the actual changes effected through these methods cannot be quantified, although some idea of the progress of modern livestock farming during 1976–82 can be derived from the data on the growth of subsidies on animal feed. The overall consequence of the policy, as we have already noted, was a change in the cropping pattern and a decline in the livestock population.

It seems reasonably clear that the basic cause of the poor performance of agriculture during the period 1976/78–1982/84, despite the rather impressive technological progress, lay in the changes engendered by the policy of promoting capitalist farming, particularly in the livestock sector, by 'hothouse' methods. In crop production, the emphasis shifted from wheat to barley and forage crops, and from high-yield crops (with a higher potential for growth) to low-yield crops (with a lower potential for growth). Furthermore, the traditional relationship of mutual dependence between crop producers and livestock holders was seriously disrupted, with modern livestock farming developing at the expense of (rather than along with) traditional livestock farming. (Livestock products account for 20–25 per cent of the total value of agricultural output in Tunisia.)

It is worth noting that the process of technological change was associated with a growing dependence of agriculture on state subsidies and imported inputs. Between 1976 and 1982, subsidies on fertilizers and animal feed grew at least ten-fold, the former from 1 million dinars to 15.8 million dinars and the latter from 3 million dinars to 32 million dinars. And these were not the only items subsidized: irrigation water, high-yield seeds, herbicides, petrol and farm machinery were all highly subsidized. Credit went almost exclusively to large farmers at very low rates of interest (significantly lower than the rate of interest on bank deposits) and much of it was in fact never repaid. The import content of inputs was high, as ammonia fertilizers, farm machinery, herbicides, cattle and animal feed were almost wholly imported and whatever was manufactured domestically also relied heavily on imported materials.

Some aspects of landholding

No data on either the pattern of land-ownership or tenancy are available. The available surveys all focus on land-holding rather than ownership. The two, of course, do not coincide because of the widespread prevalence of absentee landlordism. Our analysis, unfortunately, has to be based on landholdings and the observations must thus be regarded as tentative.

The relevant data are presented in Tables 3.6 and 3.7. From a

Table 3.6 Pattern of landholding, 1961–62, 1975–76 and 1979–80

	1961/62		1975/76[b]	1979/80	
Size *(ha)*	*% of* *holders*	*% of* *cultivable* *area*[a]	*% of* *cultivable* *area*	*% of* *holders*	*% of* *cultivable* *area*
0–5	40.8	6.4	6.0	42.1	7.2
5–10	22.4	10.6	11.8	21.4	11.5
10–50	32.6	45.4	38.6	32.3	48.1
50–100	2.6	11.7	11.8	3.0	14.6
100 +	1.6	25.8	31.8	1.2	18.6

Sources: M.B. Rowdhome, 'L'état et la paysannerie', *Mensuel*, July 1984; Ministère de l'Agriculture, *Enquête agricole de base, 1976* (Tunis, 1976).
Notes: [a] Cultivable area includes current fallows.
[b] Percentage of holders not available.

Table 3.7 Distribution of cultivated areas, 1975–76 (annual crops and tree crops), and 1979–80 (annual crops) (percentage)

	1975/76							1979/80	
Size *(ha)*	*Cereals*	*Pulses*	*Forage* *crops*	*Vege-* *tables*	*Subtotal* *(annual* *crops)*	*Fruits*	*Culti-* *vated* *area*	*Annual* *crops*	*Cereals*
0–5	4.3	12.7	4.8	18.7	5.2	8.2	6.3	7.7	6.7
5–10	10.1	17.3	15.7	28.1	11.5	14.0	12.4	11.5	11.4
10–50	36.8	40.9	31.9	36.5	36.5	42.8	38.8	42.3	43.9
50–100	12.4	8.2	12.2	10.4	12.2	9.5	11.2	12.4	13.4
100 +	36.4	21.3	35.4	6.3	34.6	25.5	31.3	26.1	24.9

Source: Ministère de l'Agriculture, *Enquête agricole de base, 1976* and *1980* (Tunis, various years).
Notes: Cultivated area = cultivable area − current fallows.

comparison of these tables, it can be seen that the pattern of landholding is roughly the same irrespective of whether we consider the distribution of cultivable area, cultivated area, or area under annual crops. Given this, there are two basic observations which can be made on the basis of Table 3.6. First, it is clear that the distribution of landholding remains extremely unequal. The bottom 40 per cent of the landholders account for only around 7 per cent of the land while the top 4 per cent account for more than 30 per cent. Second, over the long period, a slow process of decline in the average size of landholding is discernible. Between 1961–62 and 1979–80, for example, the proportion of land belonging to the largest class declined while that belonging to every other class increased.

The data in Table 3.7 indicate that the holders belonging to the size-class 10–50 hectares represent the most important and dynamic category of cultivators in Tunisia, with the most diversified cropping pattern. In contrast, the holders belonging to the first two classes seem to specialize in vegetables and pulses while those belonging to the last two classes specialize in cereals and forage crops. These patterns have seemingly remained largely unchanged during the period 1975/76–1979/80, and perhaps beyond. It thus seems clear that the observed decline in cereal cultivation is not explained either by changes in the pattern of landholding or by changes in the cropping pattern of any particular category of landholders. Data in Table 3.7 suggest that all types of landholders reduced the area under cereals, even though the reduction was sharpest for those holding 100 hectares or more.

Two other incidental but useful observations can be made at this point. First, it may be recalled (see Table 3.4) that vegetables are grown on irrigated land for the most part. Thus the distribution of the area under vegetables by the size of landholding is largely determined by the distribution of irrigated land. It can then be said that much of the irrigated land is held by those holding under 50 hectares. On the other hand, much of the fertilizer is used by the landholders belonging to the largest size-class, as can be seen from Table 3.8. This generates a certain degree of inefficiency in the use of fertilizers for the agrarian economy as a whole. In particular, it is noticeable that the landholders belonging to the size-class 10–50 hectares, who constitute the most important category of cultivators in Tunisia, actually use a lower amount of fertilizers per hectare of cropped area than most others. The reasons for this are not clear (perhaps they lie in the nature of the credit system), but it would seem to be a cause of the poor performance of agriculture in recent years. These issues deserve further investigation.

Table 3.8 Pattern of fertilizer use by size-class of holdings, 1975–76 and 1979–80

Size (ha)	Fertilizer used, 1975–76 (kg per ha)					1979–80 per cent of total
	Cereals	Pulses	Forage crops	Vegetables	% of total	
0–5	47.9	71.4	90.9	200.0	6.6	12.0
5–10	40.1	57.9	72.2	259.3	12.8	10.1
10–50	36.4	31.1	61.6	140.0	26.5	23.4
50–100	41.8	166.7	46.4	70.0	9.8	9.7
100+	70.8	65.2	100.0	216.7	44.3	44.8

Source: As for Table 3.7.

Rather dramatic changes in the distribution of livestock appear to have occurred in recent years (Table 3.9). In 1975–76 most of the livestock was held by the smallest landholders; only four years later the distribution had changed in favour of landholders holding more than 10 hectares. This change occurred in a period during which the total livestock population declined. This decline was concentrated in the class of landholders holding fewer than 10 hectares; the number of livestock held by the other categories of landholders actually increased, although this was not sufficient to prevent a decline in the overall livestock population.

Table 3.9 Livestock distribution, 1975–76 and 1979–80 (percentages)

Size (ha)	*1975–76*			*1979–80*		
	Cattle	*Sheep*	*Goats*	*Cattle*	*Sheep*	*Goats*
0–5	54.2	48.1	63.6	26.0	14.6	23.9
5–10	15.8	8.8	8.7	17.3	14.4	20.1
10–50	19.6	25.9	22.0	31.1	43.2	42.2
50–100	4.5	6.4	2.9	11.7	11.6	8.8
100+	5.9	10.8	2.8	13.9	16.2	5.0

Source: As for Table 3.7.

Thus, two simultaneous processes seem to have been in operation. The poorer landholders, who were among the traditional livestock farmers, steadily lost their viability as livestock farmers while the larger landholders increasingly adopted livestock farming as a major activity. The consequences were an overall decline of the livestock sector and an increasingly closer correspondence between the pattern of landholding and that of livestock holding. The role of the state in bringing this about is too obvious to require elaboration.

Investment in agriculture

The observed pattern of investment in agriculture provides some clues to an understanding of the growth process in agriculture and of the objectives of state policy. The observations below are based on data presented in Tables 3.10 and 3.11.

Except in the early 1970s (1972–76), agriculture's share in investment has been in broad correspondence with its share in GDP. The surprising aspect is that it was precisely during 1972–76 that agriculture experienced the fastest growth. This puzzle, however, is easily resolved if the distribution of investment by subsectors within agriculture is examined. But before doing this, we can fruitfully note a few general features of the investment process.

One remarkable aspect of investment in agriculture is the over-

Table 3.10 Investment in agriculture. 1960–61 to 1977–84 (percentages)

	1960–61	*1965–71*	*1972–76*	*1977–84*
Investment in agriculture as % of total investment	21.0[a]	20.5	13.0	16.0[b]
% of investment in agriculture				
by government	60.4	n.a.	30.7	46.5
by public enterprises	10.5	n.a.	15.5	19.9
by private farmers	29.1	n.a.	53.8	33.6
% of bank credit going to agriculture	n.a.	11.2	10.0	n.a.

Sources: M. Boudhiaf. 'Désertification de la campagne Tunisienne', (Tunis. 1986), mimeo: Ministère de l'Agriculture. *Budget économique, 1986* (Tunis. 1985); IBRD. *Tunisia: Agricultural Sector Survey* (Washington. DC. 1982).
Notes: [a] Range: 19–23.
[b] A rough estimate. derived from the fact that during 1977–81 the percentage was 13, and during 1982–84, 19.

Table 3.11 Percentage distribution of investment in agriculture by subsector, 1962–71 to 1982–85

	1962–71	*1972–76*	*1977–81*	*1982–85[a]*
Forestry/soil conservation	25.3	9.5	7.5	9.1
Irrigation	29.6	22.2	43.6	42.6
Livestock	3.9	11.5	12.0	10.9
Fruit trees	15.3	13.4	5.0	5.8
Agricultural materials and equipment	15.9	29.1	17.5	14.1
Other	10.0	14.3	14.4	17.5
Total (million dinars)	272.5	235.1	584.0	1057.6

Sources: As for Table 3.10.
Notes: [a] Provisional estimates.
'Other' includes fishing. greenhouses, cereal storage and research, studies, extension, etc.

whelming importance of public investment; even the little private investment that occurred was in essence public investment (the basis for this judgement is explained below). It is not without significance that during 1972–76 (the only period in which private investment was of major importance), the proportion of bank credit going to agriculture was at par with that during 1965–71, even though agriculture's share in total investment was much lower.

According to the government's own assessment, fewer than 20 per cent of the landholders have been beneficiaries of credit, almost all of these in the category of those holding more than 50 hectares (Ministère de l'Agriculture, *Budget Economique, 1986*, p. 82). A significant part of the credit was in the form of grants; the rate of interest was significantly lower than that paid by the banks on deposits, and purchases of materials and equipment were subsidized by the government when credit was used to make such purchases. Most importantly, no serious effort was made by the credit agencies to ensure proper repayment, so that most loans were in fact never repaid. Agricultural credit thus essentially constituted transfer payments to larger landholders. Such transfer payments accounted for much of what was recorded as private investment. The development of modern capitalist farming was thus very much a state-sponsored affair.

In such a context, public and private investments are not even acceptable substitutes. For while public investment tends to be in projects that bring long-term benefits, private investment tends to be in areas that are relevant for short-run production growth. This difference is highlighted by a comparison of the structure of investment in agriculture during the two periods: 1962–71 and 1972–76 (Table 3.11). During the period 1962–71, public investment was of prime importance and much of it went into forestry, soil conservation and irrigation development. In contrast, during 1972–76, when private investment was important, much of it went into agricultural materials and equipment. The rapid growth of agricultural output during the 1972–76 period would seem to be attributable to two types of effects: the long-term effects of the public investment undertaken during 1962–71, and the short-term effects of a shift of emphasis from public to private investment. During the period 1977–85, the relative importance of private investment again declined while that of public investment rose. Not surprisingly, the share of irrigation in investment also rose very significantly while that of agricultural materials and equipment declined. Another remarkable fact is the sharp decline in the share of the fruits subsector. These shifts explain, in part, the stagnation in agricultural production during this period.

Finally, it is worth restating that prospects of agricultural growth remained crucially dependent on the state's ability to invest. The efforts to encourage private investment were in fact scarcely concealed efforts to concentrate assets in the hands of the large landholders (those with more than 30 hectares). It is hardly surprising that these landholders failed to show much entrepreneurial dynamism.

Prices and subsidies

Agricultural products in Tunisia can be divided into two categories: those whose prices are effectively controlled by the state and those whose prices are determined by the market. Broadly speaking, prices of cereals, industrial crops, livestock products (except milk) and certain types of fruit, such as olives and wine grapes, are controlled by the state while those of most other types of fruits, pulses, vegetables, poultry products and milk are market-determined. It is important to bear these facts in mind when interpreting the observed movements in producer prices during the period under study.

Some estimates of the rates of growth of producer prices of the major agricultural products are shown in Table 3.12. Consider first the movements in the state-controlled prices. In the 1960s the prices of fruit and livestock products grew faster than those of cereals and industrial crops. During the period 1969/71–1976/78 prices of industrial crops

Table 3.12 Growth of producer prices, 1965–1969/71 to 1976/78–1982/84 (% per annum)

	1965– 1969/71	1969/71– 1976/78	1976/78– 1982/84
Cereals			
Hard wheat[a]	2.7	5.8	10.0
Soft wheat[a]	4.5	6.1	10.6
Barley[a]	2.3	8.7	10.6
Pulses	6.1	6.4	23.6
Industrial crops			
Tobacco[a]	2.4	12.6	8.1
Sugar-beet[a]	2.2	9.7	6.7
Fruit			
Olives[a]	5.2	4.1	12.3
Citrus fruits	5.9	5.6	18.0
Dates	13.9	19.2	7.1
Wine grapes[a]	17.5	8.6	9.8
Almonds	10.2	–0.3	21.1
Others	2.3	11.9	11.1
Vegetables			
Potatoes	5.5	10.2	10.9
Tomatoes	4.7	5.1	12.9
Artichokes	9.0	7.6	11.7
Peppers	9.8	8.3	12.5
Melons	1.3	11.0	13.0
Others	2.1	12.8	8.9
Livestock products			
Live animals[a]	8.1	6.8	11.6
Milk	2.5	7.5	14.0
Eggs	10.4	2.4	6.2
Fish	–1.5	11.6	17.6

Sources: Ministère de l'Agriculture, *Prix de la production des produits de l'agriculture et de la pêche, 1965–1975* and *Annuaire des statistiques agricoles, 1984* (Tunis, 1985).
Note: [a] State-controlled prices.

grew slightly faster than those of others, and from 1976–78 to 1982–84 all state-controlled prices grew more or less at similar rates. If the entire period is considered, then it cannot be said that pricing policy attempted to encourage or discourage the production of particular crops. As for the market-determined prices, it can be said that on the whole all prices moved at roughly similar rates (though there were marked fluctuations in the prices of almonds and fish) until 1976/78. From 1976/78 to 1982/84, however, the prices of pulses, some types of fruit and of fish grew faster than those of others.

A comparison between state-controlled and market-determined prices suggest that the latter grew somewhat faster than the former. A comparison between the growth of prices and that of output (Table 3.1), on the other hand, does not suggest any consistent relationship between the two. It cannot be convincingly argued that the composition of agricultural output was influenced by the movements of relative prices to any significant extent. This of course is not surprising, as prices in Tunisia are not even remotely relevant indices of profitability for a number of reasons. First, agricultural inputs were heavily subsidized. The prices of the most important current inputs were kept stable for quite long periods; the prices of others (e.g. high-yielding seeds, irrigation water, pesticides and herbicides, petrol, imported livestock, etc.) grew only very slowly. Second, much of the agricultural credit in effect constituted grants to large landholders. For these landholders, a large part of the costs of acquisition of assets as well as of the costs of current inputs was thus actually borne by the state. Third, since less than 5 per cent of the cultivated area is irrigated, agricultural production is greatly influenced by weather conditions. Indeed, it can be argued that for some crops weather conditions determine production, which determines price. Fourth, technical change has occurred at different rates for different products.

In so far as changes in the composition of agricultural output have something to do with changes in relative profitability, such profitability cannot be understood except through an integrated analysis of prices, subsidies and credit, since these affect different crops in different ways. Unfortunately, such an analysis cannot be attempted here for lack of adequate data. We can only warn against drawing simplistic conclusions about price incentives and suggest some tentative hypotheses.

The subsidies on water primarily affected production of vegetables (see Table 3.4), while the subsidies on fertilizers mainly affected production of pulses and industrial crops (pulses more than industrial crops) and of vegetables. Livestock and poultry farming benefited from subsidies on animal feed as well as on imported breeds. Credit, as noted earlier, benefited almost exclusively the large landholders and thus to an extent, livestock, poultry and cereal farming. Production of fruit and fish was least affected by subsidies. On the whole, a classification of products according to the degree of benefit from subsidies would be as follows (in decreasing order): vegetables, livestock and poultry products, pulses and industrial crops, cereals, fruit and fish. Note that advantages for livestock and poultry farming also imply advantages for cultivation of forage crops and barley.

If these observations are taken together with those on the movements of producer prices, the changes in the composition of agricultural output do not appear particularly puzzling. Pulses, vegetables and poultry farming enjoyed the benefits of both subsidies and rising demand, and hence of rising prices (which are market-determined). Cultivation of industrial crops and livestock farming profited greatly from subsidies but not so much from a rise in prices (which are state-controlled). Cultivation of fruit did not derive much benefit from subsidies; furthermore, production seems to have been much affected by weather conditions and these fluctuations in production explain the observed price trends. Cultivation of cereals and forage crops benefited significantly from subsidies and credit. In the case of wheat, the increase in yield was significant because of the introduction of high-yielding seeds. However, cultivation of barley and forage crops was attractive because livestock farming was attractive. Taken together, these factors are likely to explain the shift from wheat to barley and forage crops. Only in the case of fish does the growth of demand and hence of prices appear to have been the major stimulant for production growth.

Conclusion

The overwhelming impression that emerges from the above analysis is that the agrarian economy was greatly influenced by state policy – a complex combination of policies concerning prices, subsidies, credit and investment – even in the short run. Throughout the period, the state actively sought to reshape agriculture, first through a co-operativization movement and then through efforts to develop capitalist farming. Both programmes were financed largely by resources from outside – by foreign aid and by the quasi-rental incomes earned by the state as a result of exogenous rises in oil prices and the growth of remittances. Both programmes failed for much the same reason: no sustainable process of generation and utilization of surplus within agriculture emerged. The policy of promoting capitalist farming through massive state support cannot now be continued as oil revenue and remittances decline. On the other hand, a shift to a truly *laissez-faire* policy will have disastrous consequences for agricultural production and income distribution precisely because agricultural growth so far as been dependent on state investment, subsidies and cheap credit. Therein lies the central dilemma of agricultural policy in Tunisia today.

Chapter four

Income distribution and poverty

Our object in this chapter is to investigate in some detail the impact of the past two decades of growth on equity and poverty. The issue merits special discussion because of its intrinsic importance and because of the inadequacies of past research. In the first section we provide a profile of incomes and income distribution based on the household budget surveys of 1980 and 1985. The next section looks at the question of the incidence of poverty and the trends in poverty and income distribution since 1966, while the third section brings together the evidence on income distribution and poverty to throw some light on the growth process in Tunisia. The final section contains the conclusions. Details on the estimates of poverty lines are given in Appendix C.

Income distribution and trends

The Institut National de la Statistique (INS) of Tunisia has undertaken periodic surveys of household consumption which provide useful information on consumption levels in Tunisia. The household budget survey (HBS) of 1980 and the preliminary results of the survey of 1985 in particular contain a wealth of consumption-related information. They also give estimates of the incidence of poverty but, unfortunately, these cannot be used at their face value because of some conceptual errors in deriving the poverty line. However the consumption data in the HBSs can be used along with our estimates of the poverty line to make new estimates of the incidence of poverty, as will be seen in the second section.

Income distribution

In the context of the structural transformation undergone by the Tunisian economy, one of the aspects of income distribution that interests us is the comparison between rural and urban areas. The 1980 HBS reveals that in 1980 the rural areas, with 47.8 per cent of the total population, had 29.7 per cent of the total consumption expenditure in the country.

Although these figures might indicate a great deal of inequality, the implied rural–urban consumption gap is of the order of only 1:2.1; in other words, rural consumption was half the level of urban consumption. A similar figure is obtained for 1985, implying that rural and urban incomes increased at about equal rates between these dates: 13.3 per cent in current terms and 3.3 per cent in constant terms. Given that the propensity to consume declines with income, the *income* gap should be higher, but still of the order of only 1:3. This is not particularly high, especially when compared with other countries in Africa where, despite recent declines, the gap is even now of the order of 1:4. If in Tunisia one were to allow for the lower cost of living in rural areas, the gap would be reduced even further.

The average consumption levels in town and country are shown in Table 4.1. The highest gap – between the large cities and sparsely populated rural areas – was 2.6 in 1980 and 2.7 in 1985. The Table also shows that the average rural household was larger than the average urban household; in 1980 a rural household had 6.3 members whereas an urban household had 5.6 members. By 1985 households in both localities had declined in size, but rural households were still larger than urban households (6.0 compared with 5.4 members). The two sets of figures demonstrate the impact of urbanization on family size and the progress of family planning ideas in Tunisia.

Table 4.1 Expenditure per person and per household, 1980 and 1985 (dinars per year)

	1980		1985	
	Per person	Per household	Per person	Per household
Urban	332	1,877	619	3,352
Large cities	392	2,172	748	3,924
Medium cities	288	1,666	} 501	} 2,796
Small cities	279	1,584		
Rural	157	981	294	1,762
Principal				
agglomerations	181	1,053	390	2,248
Sparse zones	151	960	273	1,653
Total	248	1,469	471	2,665

Sources: 1980 from INS, *Enquête sur le budget et la consommation des ménages, 1980 (EB)* vol. 2, Table 1.1.2, p. 15; 1985 from INS, 'Sur les résultats définitifs de l'enquête nationale sur le budget et la consommation des ménages, 1985' ('Résultats'), INS mimeo, 16 October 1986, Table 1, p. 2.

Regional differences in consumption levels are brought out in Table 4.2. The north-east region, as might be expected (it includes Tunis and is generally the most developed region), had the highest consumption levels, followed by the south. The north-west had the lowest ranking.

Table 4.2 Per capita consumption by region, 1980 and 1985

| | *1980* | | |
	Average consumption (dinars per person per year)	*Urban- ization (%)*	*Average consump- tion 1985 (dinars per person per year)*
Northeast	324	72	450
Tunis	403	94	725
Other	239	48	n.a.
Northwest	169	25	284
Centre	208	42	n.a.
Centre-East	255	65	544
Centre-West	168	22	324
South	242	55	382
Sfax	254	59	n.a.
Others	235	53	n.a.
Total	248	52	471

Source: INS, *EB*, vol. 2, Table 1.1.3, p. 16; 'Résultats', Table 2, p. 3.

Some correlation with urbanization is shown, with the largely urban-ized north-east leading the consumption figures and the least urbanized north-west trailing. Table 4.3 carries the analysis of inequality further by looking at inter-class inequality. In both years under consideration (1980 and 1985) professionals came out at the top and agricultural workers at the bottom. Farmers themselves belonged to the next lowest class, not too far above the agricultural labourers.

Table 4.3 Consumption according to social classes, 1980 and 1985 (dinars per capita per annum)

	1980	*1985*	*Percentage change*
High-level professionals	1,017	1,500	47.5
Middle-level professionals	478	842	76.2
Employers in commerce and industry	428	907[a]	111.9
Office workers	384	768[b]	100.0
Self-employed in commerce and industry	292	456	56.2
Retired and others	258	561	117.4
Workers in industry, commerce, transport and service	231	405	75.3
Farmers	171	344	101.2
Agricultural workers	137	268	95.6
Total	248	471	90.0

Source: INS, *EB*, vol. 2, Table 1.1.5, p. 18; 'Résultats', Table 3, p. 4.
Notes: [a]In 1985 employers in services were also included here in the source.
[b]In 1985 applies to 'employed others'.

The fortunes of the different groups differed. The two top groups suffered a relative decline in their income position between 1980 and 1985, while the two bottom groups just about held their own. For groups in between the situation varied, with some notable gainers among relatively rich groups (employers and office workers) and notable losers among some poorer groups (self-employed and workers). Thus, changes in total inequality cannot be surmised from these figures; for that we need to look at frequency distributions. Table 4.4 shows the frequency distributions for rural and urban areas and for the country as a whole for 1985. The figures confirm the generally low level of consumption in rural areas as compared with urban areas: 8 per cent of the former consumed below 100 dinars per person, compared with less than 1 per cent in urban areas. At the upper end of the scale nearly 20 per cent of the urban population had consumption above 800 dinars, whereas only 3.5 per cent of the rural population had this level of consumption. This, as we said, is a reflection of relative income levels, not of relative inequality. Indeed, the calculations of Gini coefficients provided by the INS show that there was more equality in rural areas than in urban areas, the respective Gini coefficients (G) being 0.41 and 0.36.[1] With rural areas still predominantly agricultural and family-farm oriented, these sorts of figures are to be expected.

Table 4.4 Consumption distribution, 1985

Expenditure per person (dinars per year)	Population %		
	Urban	Rural	Total
0–100	0.6	8.0	4.0
100–150	3.2	16.2	9.1
150–250	13.2	31.7	21.7
250–350	18.5	18.5	18.5
350–500	22.1	13.7	18.2
500–800	22.5	8.4	16.1
800 +	19.9	3.5	12.4
Total	100.0	100.0	100.0
Memo item: Gini coefficient	0.411	0.364	0.434

Source: INS, Résultats', Tables 6 and 7, pp. 7 and 8.
Note: Population figures are in terms of per capita, not households. The same applies throughout this chapter.

Two important points have to be noted in using household budget surveys to indicate inequality. First, an HBS is designed from the narrow angle of deriving weights for a consumer price index. Thus, it records consumption expenditures and not incomes, and, being interested in deriving the cost of living affecting the majority of the people, it records consumption expenditure on the more basic items of consumption. Thus, the figures of inequality from an HBS show inequality in expenditure on items recorded in the cost of living index. Obviously, this cannot be translated into income inequality since savings are by definition excluded – as are luxury items, as well as major consumer durable goods. Now it is known that expenditure on these items increases proportionately with income, as do savings. Thus, income will always be more unequally distributed than expenditure on basic consumer goods. The second thing to note about HBSs is that in Tunisia, as in many other countries, both the very low-income groups as well as the high-income groups are generally not fully covered by such surveys. This is normally because of the difficulties of enumeration, but it may also be deliberate, as a cost of living index is designed to capture the expenditure pattern of the majority of the population on everyday items. For our purposes we note simply that income inequality will always be much higher than consumption inequality shown in a consumption survey. Although exact figures cannot be estimated, one would hazard that with consumption inequality in urban Tunisia being in the range of G = 0.4, income inequality is likely to be of the order of G = 0.6 rather than 0.5, and certainly not 0.4 as indicated by the consumption surveys. Indeed, if inequality were as low as 0.4 it may not be considered a serious problem. In rural areas, on the other hand, with a G of 0.36 for consumption, a G of 0.45 would likely be a good estimate of income inequality.

Income distribution trends

The foregoing discussion provides a background for an analysis of trends in income distribution. The accepted wisdom on this is that income distribution improved, based on the figures for the distribution of consumption obtained from the household budget surveys. These are summarized in Table 4.5.

Table 4.5 Consumption distribution according to household budget surveys (shares and Gini coefficient)

	Bottom 20%	Bottom 40%	Middle 40%	Top 20%	Top 5%	Gini coefficient
1966	6.0	17.1	36.7	46.2	18.0	0.40
1975	5.0	15.0	35.0	50.0	22.0	0.44
1980	6.0	15.0	35.0	50.0	22.0	0.43
1985	–	–	–	–	–	0.43

Sources: 1966 and 1975 from World Bank. *Tunisia: Social Aspects of Development* (Washington, DC. 1980). p. 57; 1980 from INS, *EB*. vol. 2, p. 53 (read from the graph); 1985 from INS. 'Résultats', Table 12, p. 17.

Between 1966 and 1985, consumption distribution worsened. Most of this happened in the first period between 1966 and 1975, with the top group gaining perceptibly at the expense of the bottom group. The improvement in the next period has been highlighted in official documents. This improvement is only slight; based on figures for 1980 read from a graph, the only improvement one could discern was in the share of the bottom 20 per cent. This then caused the Gini coefficient to fall from 0.44 to 0.43, which is a negligible decrease. In the next period the Gini coefficient stabilized at around 0.43. Altogether, distribution was significantly worse in 1985 than in 1966. There is an additional point about the interpretation of data in the table under discussion: they cannot even be taken as indicators of trends in consumption inequality, as they have not been corrected for class-specific price indices. Depending on how prices moved for different classes (because of differences in their consumption patterns), consumption distribution could well have worsened or improved by a great deal. Thus, if the consumption pattern of low-income groups is composed of goods whose prices increased more than the average, then distributional changes between 1975 and 1980 would be heavily regressive, and vice versa if luxury prices increased proportionately. The final point is that one cannot translate changes in consumption distribution into corresponding changes in income distribution, simply because the two are not the same. However, while we know that for each consumption distribution the income distribution is more uneven, we cannot infer from the consumption figures in Table 4.5 whether income distribution worsened. For this we have to draw on some indirect evidence.

Trends in various components of the GDP, especially the share of agriculture and wages, provide important clues on distributional changes. Table 4.6 shows this in a succinct form. Our interest for the present is to look at trends corresponding to the periods in Table 4.5. Agriculture's share of total GDP declined quite perceptibly during the period under consideration. From this, one should not conclude that income distribution worsened, since simultaneously the economy experienced massive structural transformation and agriculture's share of the labour force fell even more – from 53 per cent in 1966 to only 37 per cent in 1980. Accordingly, average agricultural income increased more than non-agricultural income, implying an improvement in the agricultural/non-agricultural income distribution.

This positive development, however, was greatly overshadowed by developments in the non-agricultural sector. Non-agricultural wages as a percentage of non-agricultural GDP fell quite substantially between 1966 and 1980, especially between 1970 and 1980, although they recovered somewhat between 1980 and 1984. Thus, there was a significant deterioration in the position of the wage earners as compared with non-wage earners between 1966 and 1980. Another way of showing this is

Table 4.6 Agricultural and non-agricultural GDP and wage share, 1966–84, selected items and years (aggregate GDP and wages in million dinars, labour force in millions, average income in dinars per year)

	1966	1970	1975	1980	1984
GDP at market prices	560	735	1,723	3,367	6,235
Agricultural GDP	100	123	315	473	821
Percentage	17.9	16.7	18.3	14.0	13.2
Non-agricultural GDP	460	612	1,408	2,894	5,414
Total labour force					
(millions)	2.257	n.a.	n.a.	3.463	4.0
Agricultural labour force	1.187	n.a.	n.a.	1.287	1.320
Percentage	52.6	n.a.	n.a.	37.2	33.0
Non-agricultural labour					
force	1.070	n.a.	n.a.	2.176	2.680
Agricultural income per					
worker	84	n.a.	n.a.	367	622
Non-agricultural income					
per worker	430	n.a.	n.a.	1,330	2,020
Ratio	5.1	n.a.	n.a.	3.6	3.2
Wage bill	227	314	608	1,153	2,288
Agricultural wage bill	30.1	38.4	68.5	n.a.	167
Wage share (%)	40.5	42.7	35.3	34.2	36.7
Non-agricultural wage share					
(= non-agricultural wage					
÷ non-agricultural GDP)	42.8	45.0	38.3	35.4	39.2
Non-agricultural/non-wage					
GDP ÷ non-agricultural					
wage	1.33	n.a.	1.61	1.89	1.55

Sources: World Bank, *Tunisia: Agricultural Sector Survey* (Washington, DC, September 29, 1982) INS, *Annuaire statistique de la Tunisie*, various years.
Note: n.a. here signifies estimates not derived.

to take the ratio of non-agricultural/non-wage GDP to non-agricultural wage – in other words, within the non-agricultural sector to compare the income of non-wage earners versus wage earners. This ratio increased throughout the period to 1980, starting at 1.33 in 1966 and ending at 1.89 in 1980. After 1980 there are signs of some improvement.

The evolution of wages is important because wage earners comprise the most important group in Tunisia. In the 1980 *Enquête Population – Emploi* they were shown to form 60 per cent of the active occupied population. Thus the evidence provided above may be taken to show that income distribution in urban Tunisia worsened, at least up to 1980. In the rural areas, too, indications are that distribution worsened. The only positive trend is that rural–urban income distribution improved, but the impact of this on overall distribution is sure to have been overshadowed by the worsening income distribution within the urban and rural areas.

The final piece of evidence we can draw on, while not quantitative, is nevertheless quite eloquent. It is based on what is known about the growth process in Tunisia. We know that the engines of growth in Tunisia were oil, tourism and remittances. The confluence of these was felt in Tunis and the coastal belt. By all accounts the interior was left behind while the coastal areas boomed. Regional inequality certainly worsened and *ipso facto* we may conclude that so did overall inter-personal inequality.

Poverty incidence and trends

A second important respect in which we differ from the accepted wisdom on Tunisia is the extent of poverty. In most writings it is held that the incidence of poverty is low (in 1985, the most recent year covered by previous analyses, it was said to be less than 8 per cent) and declining. This view derives from estimates of poverty originally made by the World Bank in 1975 using the household budget survey of that year and updated by the INS for 1980 and 1985. The problem arises because of the methodology used by the World Bank and later by the INS in estimating the poverty line. A critical review of these estimates is provided in Appendix C, and hence only the essential facts will be mentioned here.

The World Bank proposed a very simple methodology to define the poverty line. The income level of the twentieth percentile as obtained from the 1975 budget survey was taken as the threshold of poverty in both rural and urban areas. Some unknown adjustments were made to conform to the UN's Food and Agriculture Organization (FAO) norms for food consumption. The twentieth percentile level was applied to both rural and urban areas. The implications of this procedure are: (1) poverty incidence would by definition be around 20 per cent; (2) poverty lines for rural and urban areas would be different; and (3) the components of the poverty line remain unknown. The resulting estimates for 1975 were as follows: the poverty line (in terms of dinars per person per year) 79 for urban dwellers and 38 for rural dwellers; incidence of poverty 20 per cent in urban areas and 15 per cent in rural areas.

The twentieth-percentile poverty line is completely arbitrary. In the absence of significant rural-urban price differentials in Tunisia, the lower poverty line for rural areas compared with urban areas implies a different and lower standard of poverty. Thus, the figures of poverty obtained in this way cannot be compared with each other; as shown in Appendix C the D38 poverty line for rural areas is grossly underestimated and would not even buy a minimum food basket.

We have looked at poverty as an absolute concept, not a relative one as is implicit in the World Bank's methodology. Essentially, a poverty line should represent the cost of a minimum basket of goods required for survival. The more minimal the basket, the more the estimate of poverty can be advanced as representing the core of poverty. The poverty basket chosen here was indeed elementary (see Appendix C), with cereals providing 60 per cent of calories, oil 18 per cent, vegetables 11 per cent, sugar 7 per cent and meat 4 per cent. In the non-food category, assumptions were made to allow for basic clothing, housing and transport. In the choice of the food basket, we included the most basic foodstuffs; in fact, it was of a type that could be chosen for a sub-Saharan African country with a much lower level of income than Tunisia. Thus, our estimates are biased in the direction of lower poverty incidence. Estimates for rural and urban areas were made separately, using the same basket of consumer goods valued at their respective prices in the two sectors.

Poverty incidence

Our poverty lines and poverty estimates, as compared with those of the World Bank (for 1975) and the INS (for 1985), are shown in Table 4.7. For urban areas our line is not significantly different from the previous estimates; for rural areas it is totally different. We can say quite categorically that the rural poverty lines chosen by the World Bank and the INS are wrong. The mistake made by the World Bank was in adhering to their twentieth-percentile level for both rural and urban areas: for urban areas, by chance the poverty line came near to a minimum basket; for rural areas it fell far off the mark. The INS took up those poverty lines, with some modifications, and applied the consumer price index to obtain poverty lines in 1985 (and 1980; see below).

Table 4.7 Poverty lines and poverty incidence, 1975 and 1985: ILO, World Bank and INS

	1975		1985	
	ILO	*World Bank*	*ILO*	*INS*
Poverty line (dinars per person per year)				
Rural	71	38	172	95
Urban	101	79	239	190
Poverty incidence (per cent)				
Rural	43	15	31	7
Urban	34	20	16	8

Sources: ILO figures from poverty line estimates in Appendix C and consumption distribution in the relevant household budget surveys; World Bank figures from World Bank, *Tunisia: Social Aspects of Development* (Washington, DC, 1980), Tables 2 and 3; INS figures from 'Résultats', p. 17 for poverty line and Table 13, p. 18 for poverty incidence.

The ensuing poverty estimates give a false picture of poverty in rural Tunisia, as shown in the second part of Table 4.7. These estimates are obtained in the usual way, by applying the poverty lines to the consumption distribution of Table 4.4. Thus, the differences arise not from differences in the distributions used, but from differences in the poverty line. Our estimates for rural poverty are three to four times higher than those of the World Bank and the INS.[2] For urban areas, too, our estimates are significantly higher than those estimates. It is well known that poverty line estimates are subjective, because of the impossibility of a consensus on what is considered a 'reasonable' poverty basket, and thus estimates of poverty may also differ. Our objection, however, is not based on what was or was not put in the World Bank and INS poverty baskets (we actually do not know, because of the methodology used) but on how they were chosen. A conceptually flawed method of establishing the poverty line applied indiscriminately to both rural and urban areas led to an incorrect specification of the poverty line for rural areas.

Having established the poverty estimates, we now turn to examine some of the main characteristics of the poor. Table 4.8 shows the distribution of those in poverty by occupational groups in 1985. These figures are obtained by applying our estimated poverty line to income profiles as given in the household budget survey. The unemployed workers head the list in terms of the incidence of poverty, followed by farm workers,

Table 4.8 Incidence of poverty and total number in poverty by occupational groups, 1985

	Poverty incidence (%)	Numbers in poverty ('000)	Per cent of total poor
Unemployed	35.2	51	3.1
Farm workers	33.0	242	14.9
Wage earners	26.8	654	40.3
Farmers	26.6	356	21.9
Independents in commerce and handicrafts	22.7	172	10.6
Retired and others	14.2	86	5.3
Subtotal	–	1,561	96.2
Total	22.5	1,622	100.0

Source: Obtained by applying respective poverty lines to distributions in INS, 'Résultats', Table 9, p. 11. Numbers in poverty derived on the basis of Table 14, p. 19. The following poverty lines were applied: Unemployed, and retired and others, D206 (on the assumption that numbers are shared equally between rural and urban areas); independents and wage earners, D217 (on the assumption that one-third are in the rural areas); farmers and farm workers, rural poverty line, D172.

wage earners, farmers, those self-employed in commerce and handicrafts, and finally 'retired and others'. The first three categories can be said to comprise wage workers; thus one can say that wage earners constitute the largest poverty group in Tunisia. Indeed, as the figures against wage earners proper show, this group accounted for two-fifths of the poor in Tunisia in 1985. Another method of grouping is a division in terms of farm/non-farm. We could say that farmers and farm workers constituted the second largest poverty group in Tunisia, with around 37 per cent of the poor.

Finally, in Table 4.9 we look at the incidence of poverty by geographical areas. It should be noted that, lacking sufficient data in the 1985 provisional results, the table has been derived on the basis of figures given in the INS document by adjusting them by the appropriate factor to reflect our (higher) estimates of the poverty line. Thus we have adjusted the rural poverty figures by a factor of 4.45 to reflect the differential between our estimate of poverty and that of the INS, and the urban estimate by a factor of 1.85. Ideally, of course, we should apply the separate rural and urban poverty lines to rural and urban income distributions, but unfortunately these are not available by geographical locations. Given the nature of the problem at hand, the procedure we have adopted gives the right order of magnitude, as may be ascertained by applying the separate rural and urban poverty lines to the sectoral income distributions. The figures (rounded) are shown in Table 4.9.

Table 4.9 Incidence of poverty by districts and rural/urban, 1985

	Urban		Rural	
	Per cent of population	*Numbers poor ('000)*	*Per cent of population*	*Numbers poor ('000)*
District of Tunis	7	91	7	9
Northeast	18	92	26	125
Northwest	20	61	46	379
Centre west	32	78	40	321
Centre east and Sfax	10	96	19	107
South	30	183	18	80
Total	15.6	602	31.2	1020

Source: INS, 'Résultats', Table 13, p. 18, scaled up by applying the authors' estimate of poverty incidence.

The most impoverished regions are the centre west and northwest, followed by the south. Generally, poverty is higher in rural than in urban areas, except in the south where the order is reversed. In the centre west, urban poverty is at an even higher level than in the south, although it is lower than rural poverty. The western districts (northwest and centre west) contain just over one-half of the total poverty in the country.

Before moving on to discuss trends in poverty, we should underline the fact that the estimate of poverty we obtain is higher than that of the INS, especially in the rural areas, where our estimate is over four times as high. It may also be noted that the INS estimates imply both a lower incidence as well as lower absolute poverty level in rural areas as compared with urban areas. These results are anomalous and arise from the choice of the poverty line for the rural areas.

Poverty trends

Turning now to the evolution of poverty levels, let us recall that the context within which we are operating is one of rapidly increasing incomes but increasing inequality. Two outcomes are possible: first, that income distribution worsened while poverty increased; and second, that income distribution worsened, while poverty declined. The first outcome would imply a massive deterioration in bottom-level incomes; the second at least some trickle-down of growth. The Tunisian case falls in the second category, although, as we shall see, 'trickle-down' has rather a special meaning here. Several proofs can be advanced in support of this, the most important and conclusive being that based on the various household budget surveys conducted since the mid-1960s. For trends in poverty (unlike trends in income distribution) the HBSs give a correct picture as our interest is in consumption levels – the very category the HBSs attempt to capture. The procedure is to apply period-specific poverty lines to the relevant household budget survey. To render the figures comparable, an identical basket of goods has to be included in all the years. This was done, as shown in Appendix C. The resulting estimates of poverty are given in Table 4.10, together with results obtained by the World Bank, INS, and van Ginneken for 1966. The World Bank converted its twentieth-percentile line for 1975 backward to 1966 using the cost of living index, while the INS did the same with a modified World Bank line to obtain its 1980 and 1985 lines. Van Ginneken's poverty line was the then current national poverty line. Although no details are available about its contents and although in fact the earliest poverty line was very much in the nature of an aspirational target established after independence, it would seem that it hit the right order of magnitude in terms of a poverty basket. Van Ginneken then used a price differential to arrive at the rural poverty line.

All sources indicate a decline in poverty. Because of the problems with the other estimates of the poverty line, we shall describe the results obtained by our procedure. Urban poverty fell from 34 per cent in 1966

Table 4.10 Poverty line and poverty incidence, 1966–85

	Poverty line (dinar per capita per annum)		Poverty %	
	Rural	Urban	Rural	Urban
1966				
World Bank	26	55	20	27
van Ginneken	54	70	60	40
ILO	47	68	49	34
1975				
World Bank	38	79	15	20
INS	43	87	18	27
ILO	71	101	43	34
1980				
INS	60	120	14	12
ILO	109	151	42	22
1985				
INS	95	190	7	8
ILO	172	239	31	16

Sources: World Bank, *Tunisia: Social Aspects of Development* (Washington, DC, 1980), INS, *Enquête sur le budget*, vol. 2, Table 3.2.2. and 'Résultats', Table 13, p. 18; Wouter van Ginneken, *Rural and Urban Inequalities in Morocco, Tunisia and Tanzania* (Geneva, ILO, 1976); ILO: poverty line estimates in Appendix C applied against consumption distributions in the relevant household budget surveys.
Note: 1985 SMIG = 95 dinar per month, equivalent to 228 dinar per capita (five-member family) per annum.

to 16 per cent in 1985. Rural poverty also declined, although proportionately much less, from 49 per cent to 31 per cent. Combining these figures with the population distribution, we obtain the figures of total numbers in poverty shown in Table 4.11.

Table 4.11 Total number in poverty: ILO and INS, 1975, 1980 and 1985 ('000s and ratios)

	1975	1980	1985	1980:1975	1985:1980	1985:1975
ILO						
Urban	845	599	602	0.71	1.00	0.71
Rural	1,220	1,220	1,020	1.00	0.84	0.84
Total	2,065	1,819	1,622	0.88	0.89	0.79
INS						
Urban	700	393	325	–	–	–
Rural	523	430	229	–	–	–
Total	1,223	823	554	–	–	–

Sources: Obtained by applying poverty incidence estimates to estimates of total population as given in the various household budget surveys. INS figures from 'Résultats', Tables 15 and 16, p. 21.

Altogether, poverty declined by one-fifth in the decade under consideration. In the urban areas it fell by 29 per cent and in the rural areas by 16 per cent, the fall in the latter occurring between 1980 and 1985. In the urban areas, by contrast, poverty fell in the five years preceding 1980 and then stabilized in the next five years. As during this time the urban population was continuing to increase, the incidence of poverty declined, from 22 per cent to 16 per cent. In all these respects we again notice that our figures differ from those of the INS. According to the latter, poverty fell to less than one-half its level and at all times was much lower than the estimates we have derived. Quite a large part of the difference, as we have noted, arises from estimates of rural poverty, although differences also exist for urban poverty.

Table 4.12 and Figure IV.1 present further evidence to support the observation that poverty levels fell. These consist of changes in the non-agricultural minimum wage as compared with the urban poverty line. Between 1961 and 1970 the purchasing power of the minimum wage was maintained within a few percentage points of parity with 1961, although always on the low side of it, and with a notable drop in 1964. After 1970 the minimum wage began to increase – first moderately, but after 1973, and especially after 1976, quite sharply, registering a growth of 37 per cent up to 1980 and 75 per cent by 1983. After this the real value of the SMIG declined as the government was forced to adopt austerity measures in the face of the financial crisis. By 1986 the SMIG had fallen by 11 per cent as compared with three years previously. At the beginning of November 1987 the government granted a 5 per cent increase in the minimum wage (to be followed by another 5 per cent increase during 1988) but this did not suffice even to restore the erosion suffered during the previous years. The figures at the bottom of table 4.12 bring together our estimated poverty lines from Table 4.10. On the assumption that the SMIG is shared by a family of five, as implied by the population census, in 1966 the minimum wage would have bought 62 per cent of the essential needs of an average family and 72 per cent in 1975. By 1980 there was a significant improvement, with the gap falling to just 13 per cent, while by 1985, with further improvements, the minimum wage would have just about sufficed to support an average family in towns. By 1987, however, the SMIG had once again fallen below the poverty line. These figures suggest that significant improvement occurred in the position of the wage earners after 1975, which echoes trends in poverty noted in Table 4.10. Thus, between 1966 and 1975 wages improved but still remained below the poverty level; the same is true of urban poverty, which improved only slightly. Then in the next two periods wages increased substantially and it is precisely then that urban poverty began to decline perceptibly. At the same time, as we saw in Table 4.8, by 1985 around 23 per cent of wage earners

Table 4.12 SMIG (48-hour week) in current and real terms, 1961–87 (dinars per month and indices)

Year	SMIG	Cost of living	SMIG real
1961	n.a.	n.a.	105.3
1962	n.a.	n.a.	103.2
1963	n.a.	n.a.	103.2
1964	n.a.	n.a.	77.9
1965	n.a.	n.a.	90.5
1966	17.548	90	102.1
1967	17.548	n.a.	96.8
1968	17.548	n.a.	96.8
1969	17.548	n.a.	94.7
1970	17.548	100.0	100.0
1971	21.632	106.0	116.3
1972	21.632	108.0	114.1
1973	21.632	113.1	109.0
1974	27.040	117.7	130.9
1975	30.160	128.9	133.3
1976	30.160	135.8	126.6
1977	40.144	144.9	157.9
1978	44.564	153.8	165.1
1979	48.256	166.1	165.6
1980	54.704	180.1	173.1
1981	64.704	197.2	187.0
1982	85.072	224.0	218.0
1983	95.056	244.2	221.9
1984	95.056	n.a.	n.a.
1985	n.a.	286.0	n.a.
1986	105.05	302.3	198.0
1987	110.30	327.2	192.1

Urban poverty line for family of 5
1966	28.333		
1975	42.083		
1980	62.917		
1985	99.583		
1987	114.000		

Sources: 1966–83 figures from World Bank, *Tunisia: Country Economic Memorandum: Mid-term Review of the VIth Development Plan (1982–86)* (Washington, DC, October 1985); 1961–69 real SMIG from Abdel-Jalil, *L'emploi non-agricole et urbain en Tunisie* (Tunis, League of Arab States, mimeo, 1986); 1985–87 SMIG from data gathered on mission. Poverty line: ILO estimates from Appendix C; 1987 figures extrapolated from previous figures using the cost of living index.
Note: One dinar equalled approximately one US dollar in 1987.

lived in poverty. The fact that the SMIG was then just at parity with the poverty line lends support to this figure. Thus, before 1985 and especially prior to 1980, a great proportion of wage-earner families were in poverty. The increase in the minimum wage raised many of them above the poverty line but still left around one-quarter in poverty in 1985. By

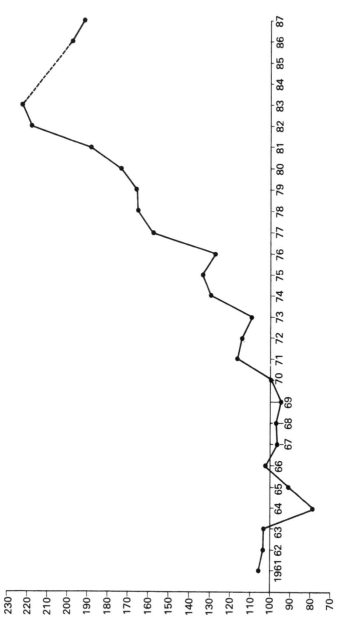

Figure 4.1 SMIG in real terms, 1961–87

1987 the position of wage earners worsened and it is quite possible that another 5 per cent may have fallen into poverty.

Growth and equity

Does Tunisia provide an example of 'redistribution with growth'? Two facts that emerge from the analysis of this chapter are that income distribution worsened but poverty levels fell. These facts have to be set against another, which provides the backdrop of this study – the strong growth in the economy. In the presence of all three trends together, economists talk of 'trickle-down' – that is, growth that is shared by all income groups. On the face of it, this is the Tunisian case; however, some significant nuances may be drawn to provide useful insights into the development process in Tunisia.

'Trickle-down' implies that growth triggered off in a certain region or among certain income groups draws other regions or income groups into the growth process. This happens through the creation of linkages as the favoured region or income groups increase their demand for products of the other regions or groups. In Tunisia this has happened in terms of the demand for food crops, although not as much as might be expected because of the switch to imported wheat (see Chapter 5). Moreover, surplus food crops themselves are grown only by a small minority of farmers. At the other extreme are families in the rural areas who own hardly any productive resources. They remained outside the growth process, properly labelled. Yet poverty declined. How did that happen? It happened because of what we might call 'trickle-up' – the process of structural transformation of the economy. Poor families (along with others) sent out their sons to the coastal towns. This raised average incomes in the rural areas in two ways: by reducing the number of mouths sharing the family pot and by augmenting family incomes through remittances. Without remittances, poverty levels would never have fallen in Tunisia because vast numbers of families lack resources in an absolute sense; if we were to look at their production levels over time we would find that they hardly increased. The same applies to many regions. These families and regions hardly shared in the growth, in terms of production. Thus, there was no 'trickle-down' in the sense of poverty groups or regions being drawn into the growth process. If their poverty levels declined, it was only because the children went out to share the boom in the towns.

Conclusion

We started this volume by noting some of the achievements of Tunisia in terms of growth and social development. We found that some of the crucial indicators of welfare had improved along with average income levels. In this chapter, we have completed the picture with a review of the trends in inequality and poverty. A number of observations may be made. First, it can be unequivocally accepted that most population groups received the benefits of growth and consequently poverty levels fell quite considerably. However, this was not because of 'trickle-down' as conventionally understood, but rather because of the process of structural transformation whereby the rural areas shed their excess population to the towns and benefited from monies remitted by relatives. Thus, the generally observed and repeated trend in poverty itself has to be qualified. Other trends from this review are unequivocally negative – and run contrary to some current views on Tunisia. These are: (1) poverty levels are much higher than hitherto believed; (2) income distribution is more unequal than hitherto believed; and (3) income distribution worsened during the two decades of growth.

It is a matter of concern that in an economy with a per capita income of over US$1,300, rather elementary poverty lines of US$350 in urban areas and US$260 in rural areas should reveal 13 per cent poverty in the former and 27 per cent in the latter. And the fact that such elementary poverty lines give such a high level of poverty confirms that a great deal of inequality exists in Tunisia. We showed that consumption itself was quite unequally distributed, with a Gini coefficient of around 0.4; as incomes are usually more unevenly distributed than consumption, we should expect a Gini coefficient of 0.6 for income distribution. Finally, by analyzing movements in the different components of the GDP, we showed that income distribution worsened, at least between 1966 and 1980 and as a trend up to 1985.

In all these respects, our findings differ from some current views. According to the Institut National de la Statistique, by 1985 poverty incidence in both rural and urban areas had declined to under 10 per cent, while income distribution was within manageable limits and had improved. We showed that estimates of poverty were flawed because of conceptual problems in deriving the poverty line. With respect to inequality, we demonstrated that the inferences drawn were wrong because they were based on consumption surveys and that the difference between consumption distribution and income distribution had not been fully appreciated.

These negative aspects of the growth process in Tunisia need to be highlighted not only because they contradict the accepted wisdom and provide new perspectives, but because taken by themselves they do constitute disturbing trends. They represent major challenges for future development in Tunisia.

Chapter five

Food consumption, food balances and subsidies

In the mid-1980s the preoccupation with the immediate financial crisis focused attention on the question of food balances. The perceived flood of cereal imports was viewed with alarm and, indeed, advanced as proof of the failure of the agricultural sector to feed the population and a contributory cause of the financial crisis engulfing the country. Even with the passing of this crisis questions still persist on the factors affecting food balances in Tunisia. It is the objective of this chapter to examine these factors by focusing on the pattern of food consumption over time. The analysis will rely mainly on the household budget surveys of 1966, 1975 and 1980. (Results dealing with consumption patterns from the 1985 survey had not been published at the time of writing.) Particular attention will be paid to the 1980 survey, both because it is the latest available and because of its wider coverage.

The discussion starts by looking at the pattern of food consumption in 1980, and shifts in the next section to the changes in consumption habits between 1965 and 1980 and factors responsible for these. This information is then used to explain the existing food balances of the country. The final section discusses the question of food subsidies.

Food consumption in 1980

Although Tunisia has by now joined the ranks of middle-income countries, diet patterns are still dominated by cereals, thus defying one of the universal 'laws' of food consumption, according to which the direct consumption of cereals declines as incomes increase.[1] In Tunisia cereals still provide around 60 per cent of total food calories, and this proportion has, if anything, increased over the years. Table 5.1 shows in a summary form the composition of diets in 1980.

Cereals provided 59 per cent of the calories ingested in 1980, and no other food group came anywhere near this category in terms of calorific contribution. In terms of expenditure, however, cereals took up only around 20 per cent of the total expenditure, the disparity

Table 5.1 Calorie and expenditure distribution of the average diet, 1980 (main food groups only shown)

Food group	Calorie contribution (%)	Expenditure composition (%)
Cereals	59.0	19.7
Fruits and vegetables	8.1	25.3
Meat	4.5	23.3
Dairy products	5.2	9.2
Sugar	6.5	4.3
Edible oil	16.2	7.2
Total	100.0 (= 2,347 calories)	100.0 (= 41.7% of total expenditure)

Source: INS, *Enquête sur le budget (EB)*, *1980, vol. 3*, Table 4.1, p. 106 for calorie contribution; and vol. 2, Table 4.11B, p. 80 for expenditure breakdown. Leguminous foods are included with fruits and vegetables, and fish with meat.

between the two proportions attesting to the hierarchy of food costs. Thus, if we were to put the calorie price of cereals at 1, the calorie price of fruit and vegetables would be around 9 and that of meat 16. This sort of hierarchy is found in almost all countries and explains why the diets of poor countries and among poorer classes are dominated by cereals. (It also provides the rationale for choosing a cereal-dominated poverty line, as in Appendix C.)

The impact of income on food consumption is shown in Table 5.2. At the bottom of the income ladder, cereals provided the bulk of calories (73 per cent), decreasing to 47 per cent at the highest income level. Following the same pattern, in the lowest-income group, cereals dominated the food expenditure, with 34 per cent of the total, falling to only 12.4 per cent for the highest group. In fact, it is only after an income level of 200–300 dinars per capita is reached that cereals lose their first place in terms of expenditure share. (Of course, they never do so in terms of calorie contribution.) The relationships between these figures are presented in Figure 5.1. While the starchy-staple calorie and expenditure ratios (proportion of calories derived from starchy foods and proportion of expenditure devoted to starchy foods) decline, the total calories from starchy foods first increase and then decline, while total expenditure on starchy foods increases. Again, these results are in conformity with dietary patterns established in other countries.

The fact that people apparently spend more to buy fewer calories tells us something significant about peoples' consumption habits. With an increase in income, the first objective is to fill the calorie gap. Hence expenditure and calories both increase, and this happens with the present sample from the first to the second income class. After that,

Table 5.2 Food consumption–income relationships, 1980

| | Income class (dinars per capita per annum) | | | | | | | |
	0–70	70–100	100–130	130–200	200–300	300–500	500+	Total
1 Cereal consumption (kg per capita per annum)	189.8	201.1	199.0	195.2	192.3	188.0	184.7	193.7
2 Calories per capita per day	1,801.0	2,146.0	2,234.0	2,308.0	2,450.0	2,559.0	2,739.0	2,347.0
3 Percentage of cereal calories	73.0	67.9	64.3	60.9	55.9	52.2	47.4	59.0
4 Percentage of food expenditure on cereals	34.1	30.3	26.6	22.7	19.2	15.8	12.3	19.7

Source: INS. *EB*: 1 from vol. 3, Table 1.5.1, p. 48 (total figure from Table 1.1.1, p. 18); 2 from vol. 3, Table 3.5, p. 99 (total from Table 3.2, p. 90); 3 from vol. 3, Table 4.5.1, p. 135 (total from Table 4.1, p. 106); 4 from vol. 2, Table 4.1.5B, p. 120; 500+ figure approximated (total from Table 4.1.1B, p. 80).

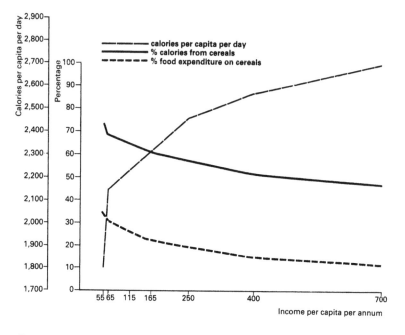

Figure 5.1 Food consumption–income relationships, 1980 (mid-point of incomes approximated)

diversification begins, as people increase their intake of non-starchy foods and superior cereals. This upgrading of cereal quality is another feature observed in all countries. In Tunisia it is neatly exhibited in terms of all the variables that affect consumption: income, urbanization and time. As incomes increase, people switch to superior cereals; with urbanization, too, diet habits change; and finally, with the passage of time people become more familiar with new foods. All these changes have important implications for food balances.

Cereal preferences

The most important cereal in Tunisia is wheat – in fact, it is virtually the only cereal consumed. But if this were the only fact known about food in Tunisia one would miss an essential part of the dynamics of food consumption in the country; the fact is that there are two types of wheat consumed in Tunisia and effectively they can be treated as different types of cereals. The traditional staple in Tunisia has been what is called 'hard wheat' or *blé dur*. Semolina and couscous are the two important derivatives of this. The other wheat consumed is called 'soft wheat' or *blé tendre*. Soft wheat consumption is almost wholly in the form of bread or *pain de boulangerie*, as it is called in the *Enquête sur le budget*.

The dynamics of food consumption in Tunisia centre around the substitution of bread for semolina and couscous. The link with food balances is provided by the fact that it is hard wheat that is predominantly grown in Tunisia, whereas soft wheat has mostly to be imported. Thus, in a very vivid way we have a demonstration of what could be called the 'African food consumption dilemma' – people want to consume what they do not produce.

Table 5.3 and Figure 5.2 show dramatically the switch from hard wheat to soft wheat associated with income; Table 5.4 and Figure 5.3 show the switch in consequence of urbanization; while Table 5.5 shows the effect of time. The income effect starts quite early, in fact, a large part occurs between the income levels of 85 and 250 dinars per capita. Now 151 dinars is the poverty line we have established for urban areas in 1980. At that level, people were already consuming 42 per cent of their cereals in the form of soft wheat. From Table 5.4 and Figure 5.3 it may be seen that the preference for soft wheat (practically all bread) is an urban phenomenon. Three-fifths of urban wheat consumption was in the form of soft wheat by 1980, whereas in the rural areas the corresponding figure was still only 18 per cent. Finally Table 5.5 shows that the consumption shift has been going on for quite some time. Between 1966 and 1985 (bringing in this time a provisional result from the 1985 budget survey), while consumption of hard wheat increased by 22 per cent, consumption of soft wheat increased by nearly three times as much.

Table 5.3 Wheat preferences: hard and soft wheat, 1980 (consumption in kg per capita per annum)

| | Income classes (dinars per capita per annum) | | | | | | | |
	0–70	70–100	100–130	130–200	200–300	300–500	500+	Total
Cereal consumption	189.8	201.1	199.0	195.2	192.3	188.0	184.7	193.7
Soft wheat	26.7	34.5	62.1	70.9	90.2	98.5	109.5	72.7
Bread	15.9	24.4	50.7	62.4	82.3	92.3	104.5	64.4
Hard wheat	146.1	151.7	124.1	114.1	88.4	95.8	59.5	107.5
Semolina	101.8	100.7	76.5	67.2	44.6	33.0	21.5	n.a.
Couscous	24.1	26.3	22.1	21.1	18.9	17.9	14.6	n.a.
Per cent soft wheat	14.1	17.2	31.2	36.3	46.9	52.4	59.3	37.5

Source: INS, *EB*, vol. 3, Table 1.5.1, p. 48; total figures from Table 1.1.1, p. 18.

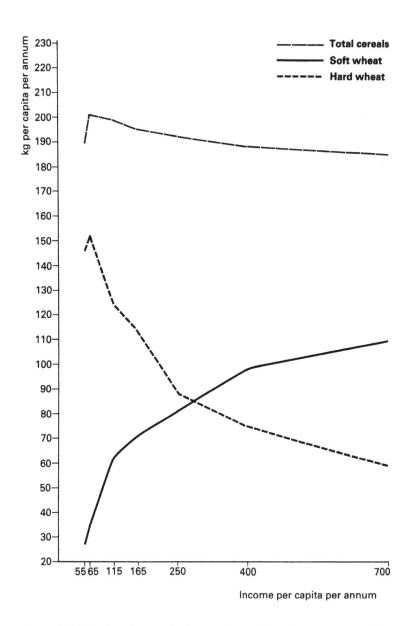

Figure 5.2 Wheat preferences by income classes (kg and income per capita per annum)

Table 5.4 Cereal preferences: rural and urban, 1980 (consumption in kg per capita per annum)

	Urban	*Rural*
Hard wheat	56.3	163.5
Semolina	20.0	108.3
Couscous	14.1	28.0
M'Hammas	3.0	8.2
Pasta	19.2	19.0
Soft wheat	103.6	38.9
Flour	4.7	12.7
Bread	98.9	26.2
Others	10.2	17.0
Total	170.1	219.4

Source: INS, *EB*, vol. 3, Table 1.2.1, p. 23.

Table 5.5 Changing wheat preferences, 1966–85 (consumption in kg per capita per annum

Total cereals	*1966*	*1975*	*1980*	*1985*
Total cereals	150	181	194	183
Hard wheat	89	108	108	102
Soft wheat	44	61	73	72

Sources: 1960–80 from Ministère de l'Agriculture, *La demande intérieure des produits alimentaires en Tunisie: Evolution et perspectives* (Tunis, May 1984), Annex 1, Table 12. Figures for 1985 from data provided by INS from provisional results of the 1985 survey. To maintain comparability, as for other years, hard wheat has been converted to grain-equivalent at the coefficient of 1.3 times bread. (From 1985, INS changed the conversion factor to 1.5.) Minor cereals have been omitted.

Absolute figures are also relevant: hard wheat consumption increased by 13 kg per capita, as compared with 18 kg for soft wheat. The figures between 1980 and 1985 indicate a possible stabilization in the consumption of soft wheat, as well as of cereals in general. This could well signify further diversification of diets away from cereals. Full results from the 1985 survey will have to be examined for this trend to be conclusively established.

What were the factors responsible for the switch to soft wheat? By now this kind of switch has occurred in so many African countries – substitution of wheat and rice for sorghum, millets, cassava, plantains, etc. – that one is tempted to argue that certain types of cereals are simply inferior foods in terms of consumer preferences and people want to get away from them as soon as they can, while others are superior foods that people prefer. Sorghum and barley are examples of inferior foods,

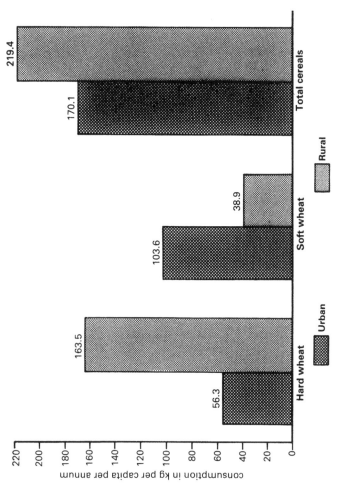

Figure 5.3. Cereal preferences rural and urban, 1980 (consumption in kg per capita per annum)
Source: Table 5.4

while rice and wheat are examples of superior foods. In Tunisia there are echoes of this, but there are also significant differences. Certainly in any hierarchy of cereals in Tunisia sorghum and barley would come at the bottom of the scale but whether bread would come at the top in preference to couscous is much disputed. There is no doubt that more bread and less couscous is consumed now but the reason is not that the Tunisians have lost their taste for the national dish. It is simply the convenience-food nature of bread: ready-made, clean, portable, and versatile. It mixes with everything that couscous, rice, maize and so on can mix with. It does not have to be eaten hot and it does not need to be prepared at home; in fact, it rarely is, and in an urbanized society that must be considered a great advantage. Couscous, on the other hand, takes a good two hours to cook, while the really traditional grain takes even longer to bring to the cooking stage. With the increased participation of women in the labour market, there are not many urban households that can find the time to prepare couscous on weekdays. In fact couscous is now very much a Sunday dish – and greatly relished. One further fact that should be noted about the rise of wheat bread and the demise of the traditional cereals as daily foods in relation to similar trends in sub-Saharan African countries is that in Tunisia, while couscous is the national dish, it has never been the 'staff of life' in the same way as maize is the staff of life in many countries of Africa (or rice in Asia). And while *pain de boulangerie* is a relatively new food, a type of bread has always existed in Tunisian diets. Thus in Tunisia, unlike countries of sub-Saharan Africa, we do not have the case of a 'superior' cereal replacing the traditional staple through the acquisition of a new taste. Rather, three or four cereals – including bread and couscous – have always been a part of the Tunisian diet, and remain so. The proportions have certainly changed, but this can be explained by the 'fast-food' nature of bakery bread.

The price regime in Tunisia has abetted the switch to soft wheat. In 1980 (and from 1975 and continuing until 1987) the price per kg of hard wheat was the same as that for soft wheat – around 96 cents per kg (Table 5.6). Because of the higher calorific content of hard wheat, the price per calorie – which provides the relevant comparison – was in favour of hard wheat, by 44 per cent. Even more relevant is the comparison between couscous, semolina and bread shown for 1987 in the second part of Table 5.6.

The 1980 calorie price differential of 44 per cent against soft wheat would easily be nullified by the convenience advantage of soft wheat (bread). By 1987 the price regime clearly favoured bread. We have a comparison of bread against couscous and semolina. The former is couscous ready to cook (but still uncooked) and the latter the grain from which couscous has to be extracted. Semolina has some price advantage as compared with bread but this would be negated by the time necessary

Table 5.6 Price regime: hard and soft wheat, 1980 and 1987 (price in millimes)

	Per kg	Per 1,000 calories
1980		
Hard wheat	95.8	27
Soft wheat[a]	97.7	39
1987		
Couscous	270	77
Semolina	155	52
Bread	167	67

Source: As for Table 5.5.
Note: [a] It is assumed that all soft wheat is bread. Calories per kg taken to be 3,500 for hard wheat (and couscous) and 2,500 for soft wheat (bread); for semolina calories set at 3,000 to allow for extraction and waste.

simply to bring it to the cooking stage. When account is taken of the cooking preparations, no advantage would be left in favour of semolina, at least in the urban areas.

The shift in the price ratios in favour of soft wheat is further elaborated with the help of Figure 5.4, in which the implied quantity ratios from Table 5.5 are also interposed. In 1961 the price of soft wheat was nearly four times that of hard wheat; by 1975 the prices had reached parity. In consequence consumption changed in favour of soft wheat, the ratio of hard wheat to soft wheat falling from 2.0 to 1.5. The astonishing fact is not that the consumption ratio declined, but that in fact it declined so much less, given the huge price changes. The continuing preference for couscous had much to do with this.

The price movements in Figure 5.4 are not 'natural' – that is, they did not come about through market forces. Between 1966 and 1975 producer prices of hard wheat were raised significantly and these prices were reflected in consumer prices. To compensate the consumers, a subsidy was granted, not on hard wheat but on soft wheat. Why this happened may easily be understood in terms of the rural-urban divide. Bread was already an urban necessity by 1975; thus to soften the blow of the increased prices for hard wheat the best commodity to subsidize was bread. Looking back from the vantage point of the late 1980s it is easy to see that a vicious circle was created: the subsidy made bread a greater necessity than ever, while the necessity dictated continuation of the subsidy. We shall look at the specific question of subsidies in a later section, but first we turn to the question of the impact of changing food habits on food balances.

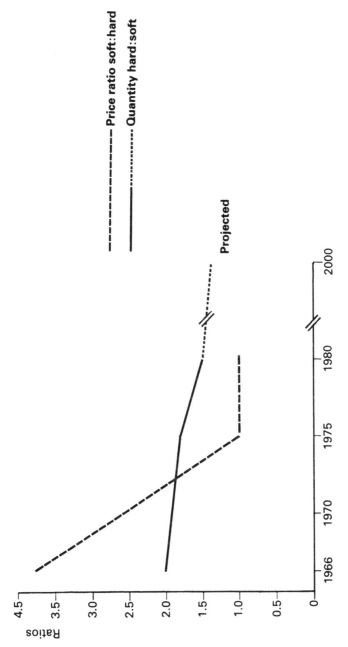

Figure 5.4 Hard and soft wheat: quantity and price ratios, 1966–80, and quantity ratio projected to 2000

Source: Table 5.5

Food balances

The question of food balances in Tunisia is addressed mostly in terms of cereal self-sufficiency, with the strong implication that Tunisia has lost its self-sufficiency in food production in recent years because of increasing cereal imports. This is true, but some qualifications need to be added. The question is also addressed in terms of the contribution of agricultural products to the balance of payments, with the implication that a strongly adverse balance in agricultural products has contributed to the overall balance-of-payments crisis. Again some qualifications are warranted. Finally, the question of self-sufficiency has to be cast within the broader perspective of the transformation of the Tunisian economy.

Table 5.7 Evolution of trade balance, selected items, 1975, 1980 and 1985 (quantity of '000 tonnes; value in million dinars)

	1975		1980		1985	
	Quantity	*Value*	*Quantity*	*Value*	*Quantity*	*Value*
Global balance						
Imports	–	572.8	–	1428.4	–	2131.4
Exports	–	345.6	–	904.8	–	1435.1
Deficit	–	227.2	–	523.6	–	696.3
Cover (%)	–	60.3	–	63.3	–	67.3
Food balance						
Imports	–	90.5	–	167.6	–	270.4
Cereals	319.4	14.6	767.5	69.5	706.9	82.9
Milk products	26.1	8.6	28.1	16.0	36.4	27.4
Soya oil	50.7	16.3	76.7	17.6	59.4	35.2
Sugar	119.4	25.0	148.3	30.5	165.5	20.5
Tea and coffee	11.0	6.2	10.2	9.1	18.6	29.6
Tobacco	5.7	4.4	4.7	7.1	6.6	18.3
Exports	–	63.4	–	61.5	–	131.8
Olive oil	41.7	31.0	37.8	22.7	46.2	42.8
Citrus	13.3	1.4	24.0	3.8	41.1	10.3
Dates	5.1	2.3	5.3	4.5	14.2	20.2
Wine	89.0	6.9	22.4	3.0	43.4	5.5
Sea food	3.7	3.6	6.7	12.4	9.4	29.8
Deficit	–	27.1	–	106.1	–	138.6
Cover (%)	–	70.1	–	36.7	–	48.7
Food imports/ total imports (%)	–	15.8	–	11.7	–	12.7
Food exports/ total exports (%)	–	18.3	–	6.8	–	9.2
Food deficit/ total deficit (%)	–	11.9	–	20.3	–	19.9

Source: Data provided by the Ministry of Agriculture, 'Evolution de la balance commerciale', computer printout.

We start with the second aspect of food balances. Table 5.7 shows statistics pertaining to financial aspects of food balances. One way of looking at these figures is to compute the special contribution of the agricultural balance-of-payments deficit to the total deficit. In 1975 agricultural exports (olive oil, citrus fruits, dates and wine) covered 70.1 per cent of agricultural imports (cereals, soya oil, sugar, tea, etc.); by 1983–85 these exports covered only 38.8 per cent.[2] Thus, the country lost a significant amount of its balance in agricultural trade. This, however, is rather a special way of looking at sufficiency. A more meaningful comparison is provided by looking at food imports in relation to total imports. Between 1975 and 1982–85 (taking an average in order to adjust for the fall in 1985) food imports increased 3.4 times in value terms while imports in general increased 3.9 times. Thus, food imports declined as a percentage of total imports, from 15.8 to 13.8 per cent. It is also worth highlighting that food imports formed only a small part of the total import bill.

Let us now look at the special place of cereals in food balances in both their financial as well as quantitative aspects. In the period 1983–85, cereals formed 40.1 per cent of total food imports by value. While this is quite a high percentage, the fact worth underlining is that cereals do not constitute the majority of imports in Tunisia, as is sometimes claimed. Certainly cereal imports have increased as a proportion of total imports – from 21.7 per cent in 1975 to around 40 per cent in 1985. The full list of imports is shown in Table 5.8. Some of the products – sugar, tea, coffee, tobacco – the country cannot produce; in others, like milk and meat, the country does not have a comparative advantage, while for soya oil the working out of comparative advantage is indicated, with Tunisia exporting the more expensive olive oil in exchange for soya oil.

Table 5.8 Food imports, 1983–85 (million dinars)

	1983	1984	1985
Total	294.8	363.0	270.4
Cereals	125.4	164.0	82.9
Milk	23.3	23.1	27.4
Meat	9.4	20.3	12.8
Animals	22.6	29.1	13.1
Soya oil	32.0	40.1	35.2
Tea and coffee	33.7	20.6	20.5
Tobacco	15.8	19.1	29.6
Potatoes	18.7	14.8	18.3
Bananas	1.9	7.2	5.7
Others	12.0	24.7	24.9

Source: As for table 5.6.

On the other side of the equation, Tunisian exports consist – apart from olive oil, which is the leader – of marine products, citrus fruits, dates and wine. Export volumes have either stagnated or not grown significantly, while prices have fared less well than import prices, contributing to the growing imbalance in agricultural trade.

Viewing cereal imports in quantitative terms in relation to production provides further insights into the question of food balances. Table 5.9 shows the relevant figures for three years between the late 1970s and 1984 and for 1985 and 1986. The importance of imports in total supply increased from 43 per cent to 54 per cent between 1977 and 1984. This could be thought of as trend; however, as the next two figures show, production is very much conditioned by the weather. Thus, in 1985 output doubled, while the next year it fell to less than one-third of its previous level. Nevertheless, a core quantity of cereals was imported even in the peak year, consisting mostly of soft wheat, testifying to its non-substitutability.

Table 5.9 Cereal supply, 1977–86, selected years ('000 tonnes)

	1977	*1979*	*1984*	*1985*	*1986*
All cereals					
Production	950[a]	950	1,023	2,104	642
Imports	717	893	1,210	732	1,312
Total supply	1,667	1,843	2,233	2,836	1,954
Import content (%)	43.0	48.5	54.2	25.8	67.1

Sources: Ministry of Agriculture, *Annuaire des statistiques agricoles*, various issues, for 1977–84. Figures for 1985 and 1986 from FAO, *Production Yearbook, 1986*, vol. 40, and *Trade Yearbook, 1986*, vol. 40.
Note: [a] Figure for 1978 used, as 1977 figure is well out of trend.

Two of the reasons for the trends up to 1984 may be deduced from Table 5.9: (1) production declined in per capita terms; and (2) demand increased faster than population growth (3.6 per cent per annum versus 2.4 per cent per annum). A third reason has to do with changes in consumption habits discussed at the beginning of this chapter. As we have seen, consumption preference has shifted towards soft wheat. Soft wheat is a commodity the country grows in very limited quantities. Thus in the three years 1977, 1979 and 1984, soft wheat provided 16, 12, and 18 per cent respectively of the total domestic production of wheat (hard wheat plus soft wheat). Demand for soft wheat, on the other hand, was 40–50 per cent of the total demand for wheat. Table 5.10 shows this for 1980, where consumption figures from the 1980 household budget survey are juxtaposed against the 1979 figures of production and imports. (1979 supply figures are used on the premiss that the previous year's supply is consumed in the current year). Soft wheat production was only 12 per cent of total wheat production (col. 6) and supplied only 13 per cent of the home demand (col. 6 ÷ col. 5). Both rural and urban areas were deficient in cereals (compare col. 5 against 6). In effect, 100 per cent of the urban demand and around 20 per cent of the rural demand for soft wheat was met through imports.

Table 5.10 Wheat demand and supply, 1980

	Consumption						
	Urban	Rural	Urban	Rural	Total	Production $(-SFW)^a$	Imports
	Kg per capita per annum		('000 tonnes)				
						('000 tonnes)	
	(1)	(2)	(3)	(4)	(5)	(6)	(7)
Hard wheat	56.3	163.5	187	498	685	450	350
Soft wheat	103.6	38.9	344	118	463	60	400
Total	–	–	–	–	1,148	510	750

Source: Consumption figures from INS, *EB*, vol. 3; production and imports from *Annuaire des statistiques agricoles, 1980*.
Note: [a] SFW = seed, feed and waste. Twenty-five per cent of production is deducted to allow for this.

Historical data extending beyond the period under survey contribute further to the question of self-sufficiency. These data have been extracted from the FAO's *Production* and *Trade Yearbooks*. We concentrate on trends in the production and imports of cereals, particularly wheat. Table 5.11 shows the relevant data, while Figure 5.5 shows the data for wheat and total production in a graphic form. We focus on three periods to derive some insights into the evolving food balances in the country: 1934–38, 1962–64 and 1982–84. Calculations of food balance sheets are shown in Table 5.12 and the resulting pattern of food consumption depicted in Figure 5.6.

Around the mid-1930s the country consumed on average 2,200 calories per person per day – that is, just about the FAO norm. Practically all the calories were produced domestically. Cereals dominated the diets, with around 60 per cent of total calories. At that time, cereals consisted almost wholly of hard wheat – eaten as couscous but also traditional bread – and barley. Oil provided around 200 calories; fruit and vegetables 50; and meat even fewer. Altogether the diet was geared towards satisfying the most basic calorific requirements of the population. By 1962–64 cereals still dominated the diets, with roughly the same contribution as before. However, some significant changes were already afoot. Almost a quarter of the cereal calories was imported in the form of soft wheat,

Table 5.11 Production and imports of wheat and total cereals, 1934–38 to 1985 selected years (all figures in '000 tonnes, except as indicated)

	Production		Imports		Total calories per person per day
	Wheat	*Total*	*Wheat*	*Total*	
1934–38	384	584	24	n.a.	n.a.
1948–52	452	689	50	n.a.	n.a.
1952	580	714	48	n.a.	n.a.
1953	n.a.	n.a.	10	n.a.	n.a.
1954	624	802	1	n.a.	n.a.
1956	n.a.	n.a.	130	n.a.	n.a.
1957	498	692	118	n.a.	n.a.
1958	536	827	27	n.a.	n.a.
1959	525	772	69	72	n.a.
1960	439	584	154	163	n.a.
1961	243	300	367	456	n.a.
1962	393	506	272	382	2,149
1963	409	626	158	172	2,134
1964	714	1,013	76	91	2,170
1965	520	712	179	235	2,306
1966	349	439	190	231	2,278
1967	330	413	336	446	2,273
1968	383	529	260	327	2,223
1969	350	444	306	392	2,151
1970	449	612	425	487	2,274
1971	600	753	229	269	2,353
1972	914	1,170	295	341	2,493
1973	690	920	216	293	2,573
1974	755	948	235	307	2,632
1975	1,035	1,268	250	319	2,649
1976	880	1,178	332	415	2,563
1977	614	777	502	684	2,620
1978	707	940	554	793	2,721
1979	680	980	707	898	2,763
1980	869	1,196	698	768	2,824
1981	963	1,263	546	960	2,815
1982	1,000	1,331	n.a.	n.a.	2,721
1983	618	956	899	1,123	2,885
1984	711	1,060	763	1,073	2,889
1985	1,400	2,124	487	707	n.a.

Sources: FAO, *Production Yearbook* and *Trade Yearbook* (Rome, FAO), various years, starting 1950. Calorie figures from FAO, food balance sheet computer printouts, FAO Agricultural Statistics Division.

all to be converted to bakery bread. Soft wheat gained at the expense of barley, traditional bread and couscous. Altogether imported calories provided around 20 per cent of total food supplies. More and more barley (and imported maize) began to be devoted to feed. The contribution of oil and 'discretionary' foods – fruit, vegetables and meat – remained about the same as in the mid-1930s period.

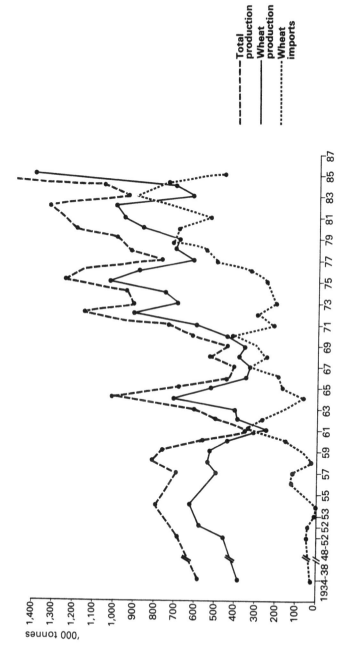

Figure 5.5 Wheat production, imports and total cereal production, 1934–38, 1948–52 and 1952–87

Table 5.12 Estimated food balance sheets for 1934–38, 1962–64 and 1982–84 (all figures in '000 tonnes, except as indicated)

	Production	Imports	Domestic supply[a]	Feed	Seed	Waste	Available food	Calories per capita per day	Of which: imported	Population ('000)
1934–38[b]										2,334
Wheat	384	24	408	–	30	30	348	988	–	
Other cereals	200	–	200	60	20	20	100	284	–	
Total	–	–	–	–	–	–	–	2,200	–	
1962–64[c]										4,256
Wheat	505	169	674	10	50	60	554	1,069	300[d]	
Other cereals	210	46	256	164	20	20	52	100	0	
Total	–	–	–	–	–	–	–	2,150	–	
1982–84[c]										6,879
Wheat	748	761	1,498	15	82	120	1,269	1,472	850	
Other cereals	352	305	683	531	55	–	75	88	0	
Sugar	–	–	–	–	–	–	–	248	217	
Fruits, vegs., pulses	–	–	–	–	–	–	–	247	4	
Meat	–	–	–	–	–	–	–	82	11	
Milk	–	–	–	–	–	–	–	88	0	
Oils and fats	–	–	–	–	–	–	–	454	293	
Total								2,832	1,375	

Source: Compiled by the authors as explained below.
Notes: [a] Production + imports + stocks – exports.
[b] Estimated. The only data are the first three figures, i.e. production and imports. Source as for Table 5.10. Allowance for seed and waste is based on normal coefficients, while that for feed is estimated. Calories converted at 3,000 per kg. Population extrapolated from 1963 figure at 2.25 per cent per annum.
[c] Estimated, based on Table 5.11 and coefficients as for 1934–38.
[d] Only wheat calories assumed to be available for human consumption; other cereal calories assumed to be used as feed.
[e] From FAO computer printout, retaining the calorie values used there. Import content estimated from underlying data in the food balance sheets. For exposition purposes, only cereal figures are shown in detail.

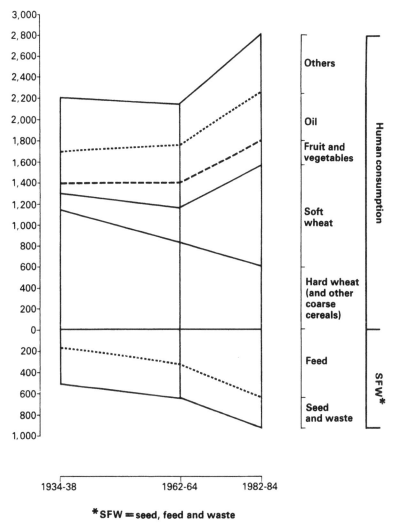

Figure 5.6 Food balances, 1934–38, 1962–64 and 1982–84
Source: Based on Table 5.12

By early 1980 the situation had changed almost beyond recognition. The country was consuming over a quarter above its calorific requirements. In addition, all the imported maize and almost all the barley was being used as feed. Thus, the country had some 3,740 calories per capita per day at its disposal but chose to give 640 (17 per cent) of these to livestock. Tunisia was moving into the category of 'big eaters' and of 'quality food', in the form of meat and livestock products. Wheat imports provided over one-half of total cereal calories and as much as 30 per cent of total consumed calories.[2] Some of the imported wheat displaced the traditional wheat, but quite a large part of the increase was simply taken as surplus food. The essential data from the 1982–84 food balance sheet also show that the country was almost wholly dependent on imports for sugar and 65 per cent dependent for oil. Altogether imports comprised just under one-half of total calorific consumption in the early 1980s.

Thus, Tunisia lost its self-sufficiency in food in a very peculiar sense: the country went beyond self-sufficiency and was helped in this by imports. Let us pause to consider the dynamics of this process. In the 1970s Tunisia's average income increased significantly. Naturally the demand for food increased, and the demand increased according to the hierarchy embodied in income elasticities. Thus, the demand increased for meat and to satisfy this demand maize was imported (to feed the stock); the demand increased for wheat bread and this was also satisfied by imports. There was an absolute inevitability about the increasing imports. No doubt the price regime contributed to some of the switch, but the switch would have happened regardless, simply because of the preference for bread. Should the government have instituted an extreme price regime to stifle the growth in the demand for bread? To what avail? To encourage self-sufficiency? Is growing, preparing and eating hard wheat to the comparative advantage of Tunisia? The calculations have not been done. In its place one is given figures of wheat imports, with the strong implication that *ipso facto* something untoward is going on and that government policy is to blame. This is simplistic economics. Given the vast increase in income that occurred in Tunisia and given the rapid transformation of the economy – urbanization and the increase in female participation – the switch to soft wheat was inevitable. And given that Tunisia's comparative advantage does not lie in growing soft wheat, the growth in imports was inevitable and justified.

Subsidies

Having looked at the impact of the bread subsidy on food balances, we now turn to its impact on equity. In Table 5.13 we first put the cereal subsidy within the context of other subsidies paid through the price equalization fund (PEF) which was established in 1970. Up to 1974 the

Table 5.13 Expenditures of the price equalization fund, 1974–83 (million dinars)

	1974	1975	1978	1983
Sugar	15.6	13.7	–	–
Cereals	–	20.0	29.3	135.5[a]
Oils	–	5.3	17.3	6.0
Meat	1.8	2.6	3.1	n.a.
Others	9.8	7.6	26.9	28.0
Total	27.2	49.2	78.4	181.0
As % of GDP	1.8	2.8	3.3	2.0

Source: World Bank, *Tunisia: Social Aspects of Development* (Washington, DC, June 1980), Table 21; and *Tunisia: Country Economic Memorandum: Mid-term Review of the VIth Development Plan (1982–86)*, (Washington, DC, October 1985), vol. I, Annex Table 11.
Note: [a] Of which subsidy on wheat amounted to D109 million, comprising D65 million as subsidy to millers and D44 million as subsidy on bread.

subsidy on sugar dominated, but after 1975 was discontinued to make room for increased subsidies on edible oils and cereals. In 1975 total subsidies amounted to 2.8 per cent of GDP; by 1978 they were 3.3 per cent and by 1983 2.0 per cent (of a much higher GDP). At this time, the subsidy on cereals clearly dominated, constituting 75 per cent of the total. Most of this (D109 million) went to wheat. Table 5.14 gives details of subsidies on some important items of consumption. Hard wheat and soft wheat (as flour) both had about the same percentage of subsidy, while bread carried somewhat more. Minor cereals such as barley and corn were even more heavily subsidized.

Who gained from the subsidies? Meat and sugar subsidies must have benefited the higher-income groups most since they are luxury commodities. Using the household budget survey of 1975, the World Bank calculated that the three highest income groups, who constituted 19.2 per cent of the population, had 44.1 per cent of the total subsidy on meat and 31.1 per cent of the subsidy on sugar.[3] Even the subsidies on edible oil were regressive in character: 34.4 per cent of subsidies went to the three highest income groups. A result that at first sight seems surprising is that even the cereal subsidy was regressive. The three bottom groups, who comprised 36.5 per cent of the population, received 24.3 per cent of the cereal subsidy, while the top three groups received 22.3 per cent. Altogether then the subsidies on the four commodities benefited the richer groups the most, with the bottom three groups receiving 22.8 per cent of the subsidies and the top three groups 28.1 per cent. Similar results are obtained from a study done by the Institut National de la Statistique on the 1980 budget survey. The percentage of food subsidies received by each successive quartile of the population was: first quartile, 12.5; second quartile, 20.8; third quartile, 29.2; and fourth quartile, 37.5. These figures

Table 5.14 Subsidies and price support in Tunisia, 1982

	Price unit	Real price	Subsidized price	Subsidy rate (%)
Hard wheat	d/q	12.229	8.000	35
Soft wheat	d/q	10.983	7.250	34
Flour	d/q	14.645	10.500	28
Semolina	d/q	18.876	12.540	34
Bread (700g)	mill	130	80	38
Barley	d/q	9.267	3.000	68
Corn	d/q	10.314	3.100	70
Soya	d/q	17.833	6.500	63
Bovine meat	d/kg	1.661	1.041	37
Poultry	d/kg	0.707	0.488	31
Eggs (unit)	mill	41	28	32
Bottled milk	m/l	230	140	39
Butter	d/kg	1.950	1.700	13
Powder sugar	m/kg	325	240	26
Mixed oil	m/l	359	300	16

Source: P. Mishalani, 'Imported inflation and imported growth: The case of Tunisia's studied postponement', in S. Griffith-Jones and Charles Harvey, *World Prices and Development* (Aldershot, Gower, 1985).
Note: d/q = dinars per quintal; mill = millimes (1,000 millimes = 1 dinar); d/kg = dinars per kilogram; m/l = millimes per litre; m/kg = millimes per kilogram; d/t = dinars per ton.

imply that higher-income groups received more subsidies per capita than low-income groups, the results obtained by the INS showing a differential of 4.7 between the highest and lowest groups for all products and 3.2 for cereals. The result for cereals, which on first reading had seemed surprising, is actually quite 'normal' given the pattern of cereal consumption; soft wheat, which receives the bulk of the cereal subsidy at the moment, is a superior cereal and consumed proportionately more by the richer groups.

However, there is also another way of looking at subsidies, which is by calculating subsidies as a percentage of income. Since expenditure on food falls with income, it is to be expected that subsidies as a proportion of income will fall. In other words, a greater part of poorer people's 'real income' consists of subsidies. In the INS document cited it has been calculated that subsidies formed 7 per cent of low-income budgets, compared with 2 per cent for high-income budgets. Cereal subsidy formed 3.5 per cent of low-income budgets, compared with 0.7 per cent of high-income budgets. Thus, the cereal subsidies benefited the lower groups proportionately more than did non-cereal food subsidies.

Finally, it is useful to compare the burden of subsidy in Tunisia with that in two neighbouring countries (Table 5.15).

Table 5.15 Food subsidies in Tunisia and neighbouring countries, 1982

	Subsidy as % of GDP		Subsidy per capital (US$)	
	Total	Wheat	Total	Wheat
Tunisia	2.8	1.9	38.92	26.41
Egypt	10.7	2.2	73.83	15.18
Morocco	3.8	1.1	33.06	9.57

Source: World Bank, *Tunisia: Country Economic Memorandum: Midterm Review of the VIth Development Plan (1982–86)*, (Washington, DC, October 1985), vol. I, Annex Table 12. The per capita subsidy figures differ from the above document as they were reconverted according to per capita GDP given in the World Bank document (Tunisia US$1,390; Egypt US$690; Morocco US$870). The percentage of the subsidy multiplied by GDP per capita then gives the subsidy per capita. The World Bank figures are slightly different: for wheat respectively US$22, US$14.60 and US$20–30.

Subsidies were far more important in Egypt than in either Tunisia or Morocco, both as a proportion of the GDP and in per capita terms. Moreover, in Tunisia, most of the subsidies went to wheat, while in Egypt a large number of items was subsidized. It is possible to argue that the question of subsidy has been rather overplayed. It is true that it entails a subsidy to high-income groups, but this is inevitable for any generalized system of subsidies, as there are very few foods whose expenditure does not increase with income. Targeting suggests itself, but often this is more inefficient and cumbersome than general subsidies. Reducing or abolishing subsidies on non-cereal foods could be more easily justified.

Conclusion

The last two decades have seen some important changes in Tunisian diet. In the urban areas soft wheat, in the form of bread, has come to dominate the traditional couscous and semolina. In the rural areas, too, there is a growing preference for soft wheat, although because of ingrained habits, lack of bakeries and lower incomes, hard wheat still dominates rural diets. The switch to soft wheat has contributed to the growing food imbalances in the country.

Government pricing policies may have been partly responsible for the growing deficit in food. One aspect of pricing policies examined here was the role of subsidies. There is no doubt that, as a result of subsidies, bread prices were kept below the market-clearing level. However, subsidies were not massive and, given the preference for bread and the great increase in income that occurred in the last two decades, the switch to bread would have happened regardless of the subsidy. Thus, subsidies alone cannot be blamed for the increasing food deficit.

Subsidies also have to be seen in terms of their impact on equity. Inevitably, generalized subsidies benefit all sections of the population and in per capita terms higher-income groups receive more subsidies than the low-income groups. However, a more relevant test is the proportion of subsidies at different income levels. On this basis, the proportion of income represented by subsidies varies inversely with income. Subsidies on non-cereal foods, on the other hand, are regressive. They are also less important in maintaining the real incomes of the low-income groups. Direct targeting of food subsidies is *prima facie* an attractive alternative, but in practice it is usually found difficult as well as costly to administer.

Chapter six

Conclusion

The main objective of this book has been to analyse the process of structural transformation that took place in the Tunisian economy over the last three decades, with particular reference to the impact on rural labour and its standard of living. A striking conclusion relates to the important shifts in employment patterns with the result that the economy changed from being predominantly rural to predominantly urban. Agriculture in particular became a minority sector in terms of employment. High rates of overall growth were achieved from the mid-1960s until the recent period, so that per capita income doubled. In the earlier part of this period, from 1965–73, growth was shared harmoniously among the various sectors of the economy. In particular, agriculture grew at about the same rate as the whole economy. After 1973 GDP continued to grow rapidly (6.0 per cent per annum) but agriculture began to falter, registering a growth rate of only 1.6 per cent per annum between 1973 and 1983. Allowing for the exodus from agriculture (agricultural labour force declined by around 7 per cent during this period), labour productivity in agriculture registered a growth of around 2 per cent per annum – perhaps more if we take account of the fact that agriculture was increasingly becoming a part-time occupation for those who remained. However, with the urban population growing at over 4 per cent per annum and with urban incomes increasing even more rapidly, the increase in agricultural output was not sufficient to meet the increasing urban demand for food and raw materials. At the same time, under the influence of rising incomes and price policy shifts, demand was switching to non-traditional cereals and luxury food. The outcome was that cereal imports increased sharply and the country became increasingly more dependent on imported food.

Thus, the fact that agricultural output failed to keep up with the growing demand constitutes one negative aspect of the growth experience in Tunisia. The economy underwent very rapid structural transformation, in which agriculture's share of employment declined rapidly, but this process of transformation was not accompanied by adequate agricultural growth, especially in the last decade. Industries and, above all, the burgeoning informal sector succeeded for the most part in absorbing the labour which had left the rural areas. The coastal areas also played their part, increasing their absorption of labour to cater for the tourist boom. Finally, precisely at this time, the neighbouring Libyan Arab Jamahiriya was going through its oil boom and this attracted some of the Tunisian labour force for employment at attractive wages. Despite these escape routes, unemployment increased in absolute terms by 38 per cent between 1975 and 1984, reaching 16 per cent of the labour force in 1984 (on a conservative estimate). This high and increasing rate of urban unemployment is another disquieting feature of Tunisia's growth experience. The trickle-down to the poor groups during the boom years has shown signs of drying up. In particular, the possibilities open to informal sector employees to share in the boom have been rapidly decreasing since the early 1980s.

A further probe into the factors affecting agricultural productivity reveals other weaknesses of the growth process. Progress in agriculture was very much dependent on the bounties of the state, which had become the distributor of subsidized (and often imported) inputs: machinery, credit, fertilizers. Most of these went to the large-scale farmers who, however, failed to become dynamic entrepreneurs. The small farmers, especially in the traditional sector, were constrained by their low levels of technology and by their inability to secure adequate credit. The major consequence of these trends was that even during the period of high productivity growth, the agricultural sector failed to generate any sustainable process of internal accumulation.

On the whole, the impression one has of the Tunisian experience is that of growth based on windfall gains. The economy was propelled forward – and at a very rapid pace – by the three engines of growth of the economy: tourism, remittances and oil. In the post-1973 period the impetus from all these sources increased tremendously; tourism benefited from the infrastructure laid down to attract it, and remittances and oil receipts from the oil price boom. The internal forces of growth increasingly took a back seat to the external boom; at least in the agricultural sector, the internal dynamism for growth was probably even stultified. In the urban areas new industries were created, but most owed their existence to heavy protectionism. All this could be masked and even tolerated while the boom lasted. The country could afford to set aside some of its abundant foreign exchange to import food or subsidize import-dependent local industries. Once the boom ended, the whole basis of subsidies to agriculture and import-intensive industrialization began to weaken. The economy was perhaps left in an even more vulnerable position than before.

In terms of the welfare of the Tunisian people, there is little doubt that a significant improvement was achieved. Evidence suggests that there was some trickle-down of growth, especially in the period 1973–83. Poverty levels declined perceptibly – by half in the urban areas and one-third in the rural areas. Yet it must be noted that in a country with a per capita income of over US$1,300, 31 per cent of the rural population and 16 per cent of the urban population could still be counted to be in poverty in 1985. In a relative sense, too, poverty increased over time. Here we refer to the state of income distribution in the country. Income distribution worsened during the boom in Tunisia. Statistics to show this are scarce, but certain pointers can be established. Between 1966 and 1980 the share of wages in the GDP declined from 41 per cent to 34 per cent, and the ratio of non-agricultural/non-wage GDP to non-agricultural wage increased from 1.33 to 1.89. These figures imply a redistribution of income in favour of the non-wage groups: the traders, industrialists and, to a lesser extent, the informal sector operators.

Before moving on to the second part of this chapter, we may usefully summarize our account of the Tunisian experience of the last two decades by the following three propositions:

1 GDP grew, but income distribution worsened.
2 Poverty declined, but poverty levels were higher than originally thought.
3 The economy underwent a structural transformation, but the structures created were not self-sustaining.

The last is the most important conclusion of the study: despite an impressive performance during the decade 1975–84 by developing country standards, Tunisia did not manage to achieve an accumulation pattern that guaranteed sustained economic growth without rental income from abroad. 'Growth without development' is a charge normally levelled against this experience, but that would be too harsh a judgement for Tunisia, as some genuine employment was created in the urban sector and growth did trickle down to the masses. Precisely because development in the agricultural and industrial sectors ran into difficulties when the external stimuli disappeared, the economy found itself in a crisis.

Clearly, adjustment has to be made to the new economic environment characterized by the reduced availability of quasi-rental incomes. Two major imperatives have to be faced: (1) the need to make up for the loss in foreign exchange as a result of the reduction in external resources; and (2) the need to increase productive employment in both rural and urban areas. So far the urban informal sector and external migration have acted as safety valves for the increasing labour force, but obviously this cannot continue forever. The VIIth Development Plan proposes a strategy based on export promotion. In the long run this perhaps may be the correct policy, but the solution of the immediate problems facing Tunisia requires a somewhat different approach. While certain stabilization measures are needed to tackle the short-run problems of external imbalances, sooner or later attention will have to turn to the more basic problem of the imbalance in the economic structures that emerged in the last two decades. In line with our thesis we are convinced of the necessity of such long-run measures to correct the basic imbalances in the economy.

Looking towards the future, it appears that Tunisia has little option but to consider a two-pronged strategy of development that aims at the promotion of non-traditional exports and the increase in food supplies. Such a strategy would have two principal components. First, possibilities for developing non-traditional exports could be explored; this indeed is an idea proposed in the VIIth Development Plan. If exports of non-traditional items could be generated or increased so as to make up for the loss of quasi-rental incomes, the government would have the resources both to survive the present crisis and to pursue a new investment strategy. This strategy would need to be much more oriented towards generating a sustainable internal process of accumulation than has been the case in the past, but this would be a feasible option. The problem is that, given the present set of exports, it is difficult to envisage the generation of new exports to the extent required, except perhaps in the very long run. If reliance is placed on foreign aid or foreign private investment, then the government can hardly control the investment pattern. Moreover, this strategy will only postpone the fundamental problem which will eventually manifest itself in the form of a debt crisis.

The second component relates to food security. Tunisia needs to embark on a process of reorientation of the development strategy, from one primarily dependent on foreign resources to one reliant on internal accumulation. In such a strategy agriculture will have to play a much more important role than hitherto; it will have to be viewed as a provider of food and raw materials to the urban economy as well as a generator of surplus needed for its own growth. This will help reduce import needs, and an accompanying policy of curbing luxury consumption in urban areas will go a long way towards solving the balance–of–payments problem. In this perspective, agriculture's contribution to employment generation and GDP will remain important and perhaps even increase in the foreseeable future.

A central requirement of this strategy is a significant alteration of the property relations in agriculture. One possibility is to make the middle category of landholders (i.e. those holding between 10 and 50 hectares) the basic target group for state policy. At present, this group accounts for about 60 per cent of the cultivated area and livestock. Improvements in the position of this group therefore will have a significant impact on agricultural production. Furthermore, efforts can be simultaneously made to carry out land reforms so as to transfer surplus land from those holding more than 50 hectares to those holding fewer than 10 hectares. This strategy will both reduce landlessness and expand the size of the group of middle farmers. The reason for concentrating on the medium-size holdings is that they constitute the most dynamic group among Tunisia's farmers, one capable of producing a surplus and investing it in the agricultural sector. State support in the form of direct investment, input subsidies and credit will still be needed, but the required level of support per unit of output would be lower than at present, as middle farmers are generally more efficient than large farmers. Such a strategy, moreover, will stimulate employment in agriculture and discourage mechanization. Within such a context it will be both necessary and desirable to shift the emphasis from labour-saving to land-saving technological changes in crop production. Agriculture's dependence on imported inputs can thus be reduced while productivity growth is sustained. In the livestock sector, too, efforts can be made to increase the viability of traditional livestock farming.

Two qualifications to the above arguments should be noted. First, even if this strategy is adopted, foreign aid may still be needed to tide over the short-term difficulties. However, this aid will help eliminate the need for aid beyond the short run. Second, the strategy is not easy to implement, since it requires a strong commitment on the part of the government in the face of the opposition that might emerge to the attempts to change the existing agrarian structure. However, genuine development always involves systematic changes and difficult choices. This, in essence, is the challenge facing Tunisia now, especially after the change of government of 1987.

Appendix A: Labour migration from Tunisia

Tunisia has a long history of labour migration. After independence some 200,000 people left Tunisia. These were mainly French colons who departed after 1956, as well as Algerians and Libyans who used to live in Tunisia and were attracted home after the discovery of oil. By the 1960s, a large wave of emigration began from Tunisia towards Europe in general and France in particular. One of the main reasons for this was the forced introduction of co-operatives, especially for smallholders. Young male members of peasant households left the country to seek employment in the then booming economies of Europe. By the 1980s, as many Tunisian workers began to return from Europe as a result of recession, a new wave of emigration began towards the Libyan Arab Jamahiriya and to a lesser extent towards the Arab oil-producing countries of the Gulf.

Despite this long history of emigration, the exact number of migrant labour is not known. Various figures are presented ranging from 180,000 to 309,000. For instance, the *Rapport du Comité Technique de l'Emigration*[1] estimated that the Tunisian community in Europe amounted to 309,549 people in 1984, as shown in Table A.1.

These figures appear to be exaggerated. For instance, according to the sources of the French Ministry of Interior, the Tunisian community was estimated at 212,900 by the end of 1982. Moreover, the French population census (7.3.1982) estimated the number of Tunisians living in France at 189,400. This is obviously a lower figure than that mentioned by the *Rapport*, which relies on the consular records as a source of its information. Our guess is that the data provided by the *Rapport* overestimate the size of migrant labour as a result of double counting. Many short-term migrants register more than once at the consulate and this explains the resulting high figure.

The *Rapport* estimated the number of Tunisian workers in other Arab countries as being 155,190 in 1984–85, distributed as in Table A.2. If we add these numbers together, the total migrant labour will amount to 464,739 according to the *Rapport*, which seems to us an overestimate.

Table A.1 Estimate of Tunisian migrants in Europe, 1984

	Total	% males[b]	% females[b]	% children less than 16 years old
France	258,378	52.0	16.2	31.8
Fed. Rep. of Germany	23,644	46.6	22.0	31.4
Italy	16,000[a]	n.a.	n.a.	n.a.
Belgium	9,527	50.0	22.1	27.9
Austria	2,000	n.a.	n.a.	n.a.
Total	309,549	–	–	–

Source: Ministère des Affaires Sociales, *Rapport du Comité Technique de l'Emigration* (Tunis, June 1985)
Notes: [a]Including 3,400 legal emigrants and 12,600 illegal emigrants.
[b]Does not include children below 16 years.

Table A.2 Estimate of Tunisian migrants in Arab countries, c. 1985

	Number of workers
Libyan Arab Jamahiriya	100,000[a]
Algeria	41,136[b]
Iraq	1,118
Saudi Arabia	8,500[c]
Kuwait	2,000[d]
United Arab Emirates	676[d]
Oman	600
Qatar	810
Jordan	350
Total	155,190

Source: As for Table A.1.
Notes: [a] Refers to the first half of 1985. 32,000 Tunisian workers were dismissed in September 1985.
[b] Refers to 1984.
[c] Refers to beginning of 1985.
[d] Refers to 1984.

The above figures include Tunisian workers and their dependants. For instance, according to the *Rapport*, the members of the labour force living in Europe amounted to 136,244, or 44 per cent of the total number of Tunisians living in Europe. Adding to this the labour force in the Arab countries, the total Tunisian labour force abroad would come to some 180,000 at the end of 1985, after allowing for the waves of return migration from both Europe and the Libyan Arab Jamahiriya. On this basis migrant labour represented around 10 per cent of the total labour force in 1985.

This sizeable migration has affected the labour markets to a great extent. In fact, migration since the 1960s has provided an important source of employment opportunities for surplus labour in Tunisia. Various development plans have relied on emigration as an important source of employment. Despite a pronounced decline in the ratio of migrant labour, migration was still expected to provide 'jobs' for some 7.5 per cent of the labour supply. This points to the serious consequences of return migration, which began by the early 1980s. It has been estimated that between 1979 and 1984 some 34,310 Tunisians returned from abroad, or roughly 5,000 returnees per year.[2] Given the high rates of unemployment in Tunisia, this represents an obvious added burden on the Tunisian economy.

Appendix B: Estimates of unemployment

Official estimates of unemployment in Tunisia tend to underestimate the number of unemployed. (Such estimates, based on the 1975 and 1984 population census, amounted to 12.9 and 11.5 per cent respectively for the country as a whole.) The main reason is that the definition of unemployment used by the census suffers from a number of conceptual shortcomings. In the two censuses, unemployment figures included those who declared themselves unemployed in the age groups of 18–59 years. This left out the age groups of 15–17 and those above 60 who were looking for work but could not find it. Youth unemployment among the age group 15–17 is extremely high. In fact, this is one of the problems the government has been trying to tackle for a number of years. The 1984 population census provides the following arguments for excluding these age groups: (1) the 15–17 group can be excluded on the grounds that they are too young to be in the labour market, the minimum age for employment in the formal sector being 18;[1] (2) those over 60 are too old for employment and qualify for social services.[2] Tables B.1 and B.2 provide the results of our attempt to correct for the omission of these two age groups. The estimate of unemployment increases from 10.6 to 15.7 per cent in 1975 and from 11.5 to 16.4 per cent in 1984.

The rate of unemployment should be even higher than the 16.4 per cent just derived. If we take the results of the 1984 population census, for instance, we can identify two other sources for underestimating the unemployment ratio. First, the census classified as employed in the rural areas all those who reported that they had land.[3] Given the nature of the rural Tunisian society, where land is held by the clan or extended families, many people could report that they have land in the sense that they have some claim on the family land. This should not exclude them from being counted as unemployed, as it affects 90,000 members of the labour force. Second, the census classified a large number of women as housewives who are involved in handicraft production, especially carpets, clothing and leather goods. They are usually involved in

Tunisia

Table B.1 Structure of the labour force, 1975

	Urban	%	Rural	%	Total	
1 Labour force 15–60	828,920	51.1	792,900	48.9	1,621,820	100.0
2 Employed 15–60	718,700	52.6	647,820	47.4	1,366,520	100.0
3 Unemployed 18–59	72,640	–	99,700	–	172,340	–
4 Active but unemployed <18 and >60	37,580	–	45,380	–	82,960	–
5 Adjusted unemployed (3 + 4)	110,220	43.2	145,080	56.8	255,300	100.0
6 Unemployment rate:						
Not adjusted (3 ÷ 1)[a]		12.6		8.8		10.6
Adjusted (5 ÷ 1)[b]		13.3		18.3		15.7

Source: Recensement général de la population et des logements, 8 May 1975, vol. 51, *Caractéristiques économiques*.
Notes: [a] Population census figures.
[b] Adjusted estimates by adding to the unemployment figures reported by the census, those in the age groups 15–17 and 60 and over who seek employment, since the census excludes them. The 1984 census reports a category of *actifs potentiels* which includes: (1) those who were unemployed in the week before census and are preparing for self-employment; and (2) those between 15–17 and over 60 who look for work.

Table B.2 Structure of the labour force, 1984

	Urban	%	Rural	%	Total	%
1 Labour force 15–60	1,166,720	54.6	970,490	45.4	2,137,210	100.0
2 Employed 15–60	991,790	55.5	794,630	44.5	1,786,420	100.0
3 Unemployed 18–59	126,350	–	118,260	–	245,240 (630 invalid)	–
4 Active but unemployed <18 and >60	48,580	–	57,600	–	105,550	–
5 Adjusted unemployed (3 + 4)	174,930	49.9	175,860	50.1	350,790	100.0
6 Unemployment rate:						
Not adjusted (3 ÷ 1)[a]		10.8		12.2		11.5
Adjusted (5 ÷ 1)[b]		15.0		18.1		16.4

Source: INS, *Recensement général de la population et de l'habitat*, 1984, vol. 5.
Notes: As for Table B.1.

production through a 'putting-out' system where they produce their goods at home and market it through middlemen. The census classified all these as housewives, excluding them from the labour force.

If we were to correct for the three sources of bias in the official estimate of unemployment, the unemployment ratio might increase to 25 to 30 per cent of the labour force.

Appendix C: Poverty lines in Tunisia: A critical review

There is a long tradition in Tunisia of setting poverty lines and measuring the progress in eradicating poverty. Inevitably there is also a lively interest in these estimates, and indeed much controversy surrounds them. The objective of this appendix is to contribute to this debate.

We start with the observation that poverty estimates are as good as the poverty line on which they are based – and a poverty line is as good as the basket of goods it embodies. The problem with the poverty lines so far established in Tunisia is that insufficient detail is available on their composition to facilitate a judgement on their suitability. However, elementary checks on the data are possible and these indicate that the extant poverty lines in Tunisia are incorrect. (Hence the poverty estimates are also incorrect.) One of the poverty lines in existence was established soon after independence and seems simply to be in the nature of a goal to be reached. The other, originally conceived by the World Bank, has somewhat more methodological content to it but as it was based on an arbitrary cut-off point, the poverty line was placed too low. Later studies by the Institut National de la Statistique (INS) have carried this poverty line forward. According to these estimates, there has been a fall in poverty levels over time, which we found to be correct; but at each stage poverty was shown to be much lower than it should be.

We start by examining in detail the poverty line established around 1980 by the World Bank for 1975, now almost the official poverty line in Tunisia. After this we shall look at the 'traditional' poverty line set by the Tunisian authorities around 1960, now partially defunct, although still mentioned in the plan documents. Then we shall examine the derivation of the poverty lines used by the INS. Finally we shall present our own estimates of the poverty line.

The World Bank estimates of poverty lines

The World Bank's poverty lines were established by a mission around

1980.[1] They were said to be

> an absolute poverty measure based on the local cost of minimum nutritional and non-food requirements (clothing, shelter). . . . The absolute poor are those people below a certain income level whose basic human needs cannot be readily fulfilled.
>
> (p. 9)

The poverty thresholds were derived 'using a methodology generally followed by the World Bank':

> This methodology calculates the nutritional element of the poverty line based on a typical food basket of the low income groups, e.g. that of the 20th percentile of the household expenditure distribution. The calorie content and the cost of this diet are then determined and adjusted to FAO standards.
>
> (p. 10)

The idea appears to be to take the food expenditure of the twentieth-percentile class and modify it to conform to 'FAO standards' of calorie requirement – around 2,200 calories per day. This does not constitute a sufficient basis for setting a poverty line, since: (1) the food expenditure of the twentieth percentile group may not be sufficient to meet the FAO standard; and (2) even if it were adjusted, 2,200 calories may be obtained from any combination of sources. If topping up is being done one might as well admit that the poverty line is derived independently of the household budget survey, a procedure that is inevitable and was evidently followed in this case. No further explanation is provided on the calculations done except to give a table. This table is reproduced here in its entirety as Table C.1. Clearly, without any accompanying explanation it is not too helpful in understanding the Bank's methodology. We shall thus attempt to present the assumptions on which the estimates were made.

The World Bank used its methodology to derive the 1975 line and worked backwards to the 1966 line. We shall therefore first look at the 1975 figures. As a start, it should be noted that food expenditure figures by percentile classes are not available in the published household budget survey document. (Perhaps the Bank was made privy to these.) Second, contrary to its assertion about the twentieth-percentile methodology, the rural poverty line was not put at the twentieth-percentile level, for otherwise we should obtain 20 per cent poverty in rural areas.

In fact, what we notice is that the rural poverty line is set at only 48 per cent of the urban poverty line. Now one of the cardinal rules about making poverty comparisons – whether across groups or over time – is that the same poverty line must be chosen as a bench mark. To express this idea in physical terms, it would obviously be incorrect to set the poverty line for urban areas at 2,300 calories and for the rural

Table C.1 Absolute poverty threshold according to World Bank (in dinars)

	1966		1975	
	Urban	Rural	Urban	Rural
1 Food expenditure of the 20th percentile (per person/month)	2.385	1.629	3.416	2.333
2 Adjusted expenditure to meet nutritional requirements	2.273	1.302	3.414	1.956
3 Cost of nutritionally adequate cereal-based diet (per person/month)	1.298	0.878	1.950	1.320
4 Non-food expenditures of 20th percentile (per person/month)	2.352	1.063	3.166	1.416
5 Adjusted non-food expenditure	2.350	0.892	3.163	1.187
6 Absolute poverty line (2) + (5)				
(i) per person/month	4.623	2.194	6.577	3.143
(ii) per person/year	55.476	26.328	78.924	37.716
Memo items				
Poverty incidence (%)	27	20	20	15
Cost of living index	100		143.2	

Source: World Bank, *Tunisia: Social Aspects of Development* (Washington, DC, 1980), Tables 2 and 3. It is stated that these were 'Bank staff calculations based on data reported in the INS consumption surveys of 1966 and 1975'.
Note: Poverty figures for 1966 given in Table 3 and included above are incorrect. As may be ascertained from the budget survey as well as World Bank's own appendix Table 10, 22 per cent of the urban population and 17 per cent of the rural population fell below the stipulated poverty line.

areas at 1,800 calories. The ensuing poverty figures would be incomparable. This is effectively what has happened here. If prices between rural and urban areas differed by one-half, one could say that the rural poverty basket chosen was comparable to the urban basket; if prices are equal then the rural basket would be only one-half that of the urban basket. The Tunisian case falls in between: in keeping with the implicit assumption of the World Bank calculations, there is a price differential in total in favour of rural areas – but not on all items and not to the extent implied by these figures. A favourable price differential exists for some foods produced in rural areas, but not for all. For 'town' foods such as tea, coffee and so on, or for non-food items, most of which originate in towns, the price differential should go against the rural areas. This is to be expected since distances are small in Tunisia. Even for food, the price differential is not of the extent implied in the World Bank's figures. A price differential exists for meat and fruit and vegetables, but not for cereals, whether in the raw or processed form, since according to Tunisian practice, prices are controlled and set equal for all areas. The price differential implied by line 2 is 75 per cent. As shown in Table C.4, the price differential on an equal basis – that is, setting rural and urban diets at the same qualitative level – was only 16 per cent. Thus, if the World Bank gets a 75 per cent cost difference in the two baskets, that can only be because its urban basket has a relatively greater proportion of the more expensive items, implying that the two food baskets were not equal. When we come to the non-food basket, we find the cost differential even greater: 2.7 in favour of urban areas. For the 'equal-baskets' rule to apply, prices of non-food items – clothing, fuel, cleaning material and so on – in urban areas have to be 2.7 higher than in rural areas. In fact, the opposite is the case. If we put a modest differential of 1.2 against the rural areas, the figures in line 5 imply that the rural non-food basket was only 32 per cent of the urban basket. Even if we say that prices in rural and urban areas were equal, the rural non-food basket was set at only 45 per cent of its urban equivalent. Of course, there are 'non-discretionary' expenditures in towns that have to be included in the poverty basket, which do not figure to any great extent in rural areas – transport in particular. And some items do cost more in urban areas because of quality differences – housing, prominently – but even then the figures we have imply a vast differential in the non-food baskets: at least 1:1.8. Altogether, the differential in the total basket could well be of the order of 1:1.6. Exact figures would be meaningless; the point that should be made is that the urban and rural baskets are not equal and therefore not comparable.

What about the absolute level of the poverty lines? It is difficult, as we have just seen, to say anything concrete about the non-food component, but something insightful can be said about the food component. We start with the figures in line 3 of Table C.1 and explain their significance and derivation. They show the cost of a food basket in which all the calories come from cereals – that is a *100 per cent* cereal-based diet. (The World Bank does not recommend this diet.) We find from the 1975 household budget survey[2] that in the rural areas an expenditure of D12.322 per year (p. 233) bought the equivalent of 1,663 calories per day (p. 397). Thus, to obtain 2,200 calories from cereals, D1.36 would be needed per month, which is the sort of figure in line 3. Similarly, in the urban areas D13.992 was spent during the year to obtain 1,223 cereal calories per day. Thus, D2.097 per month would be needed for a 100 per cent cereal diet – which is close enough to the relevant figure in line 3.

Now in going from line 3 to line 2 of Table C.1 we notice that the World Bank allows a calorie-price differential of 1.75 between the 100 per cent diet and the unspecified cereal/non-cereal diet in the urban areas. What sort of diet could this have been? Non-cereal foods are of course expensive sources of calories – to be precise, 4.4 times as expensive as cereals according to the household budget survey in question.[3] From the figures given in note 3 it may be calculated that a 50:50 cereal/non-cereal food basket[4] would have cost D5.511 per month, or some 61 per cent above the cost shown in line 2. A diet with a monthly cost of D3.414 would consist of 80 per cent of calories from cereals and only 20 per cent from non-cereals. Now, as the World Bank had itself noted, 'Although, theoretically, a diet based on cereals, with some supplementation to provide adequate balance may be nutritionally adequate, this may not be acceptable to the population'. If we were dealing with a very poor country – say, one with a per capita income of US$200 – we may well set an 80:20 cereal/non-cereal diet as the basic food poverty line. In Tunisia we are dealing with a country with a per capita income of US$800 in 1975. Put differently, the urban food poverty line chosen was only 55 per cent of the average food expenditure in urban areas in 1975 (D74.30 per annum). At the average expenditure level, cereals still provided one-half of total calories so that the average expenditure level was still certainly a 'no-frills' situation. To put the food basket at only 55 per cent of this with an 80:20 cereal/non-cereal breakdown takes us into the realm of austerity.

Thus, two conclusions emerge about the World Bank's 1975 poverty lines. First, poverty lines for rural and urban areas were different, with the former appreciably lower than the latter. Second, food poverty lines chosen were much too austere for a country such as Tunisia in 1975.

The 1966 poverty lines

Despite its proposed 'methodology', the World Bank's 1966 poverty lines were simply interpolated from the 1975 estimates. This conclusion is reached when we notice that the relationship between the various urban and rural components has the same order of magnitude in the two years.[5] This implies that the authors believe that the same rural/urban price ratios existed in the two years, which is obviously not the case. Our conclusion is further confirmed when we find that the poverty line chosen in 1966 is 70 per cent of the 1975 line, which is the ratio of the cost of living index in the two years. Thus, we infer that the poverty line in 1966 is obtained simply by applying the cost of living index to the 1975 poverty line. Finally, when we return to the household budget survey in question, we find that the poverty lines chosen bear no relationship to the income figures there.

Let us investigate the feasibility of the poverty lines chosen by examining line 3 of Table C.1 against the cost of a 100 per cent cereal diet (Table C.2). The price of cereals in rural areas was 73 m per kilogram, or 21 m per 1,000 calories. At this price a 100 per cent cereal diet providing 2,200 calories per day would cost D1.371 per month. This is some

Table C.2 Cost of a cereal diet, 1966

	Rural	Urban
Cereal price (millimes/kg)	73	107
Calories/kg[a]	3,500	3,500
Monthly cost for 2,200		
calories (in dinars)	1.371	2.025
Memo items		
World Bank cost (in dinars)	0.878	1.298
World Bank full diet (in dinars)	1.302	2.273

Source: INS, *La consommation et les dépenses des ménages en Tunisie, 1965–1968*. Price from Tables A2-1 and A2-2. Calories per kg as indicated below.
Note: [a] This figure is generated by the survey (see pp. 151 and 402). Note that the figure is different from that generated by the 1975 survey – 2,900 calories/kg. The difference can be attributed to an increase in the proportion of bread, which has a lower calorie content than cereal grains.
Calculations based on unrounded figures.

56 per cent above the cost entered by the Bank (D0.878). The same differential is obtained for the urban areas. The World Bank's calculations here are simply incorrect. For some reason the Bank does not seem to have gone back to the source to derive the ruling price of cereals and then make the sort of calculations we have made (or the Bank itself made for 1975). It simply applied the cost of living index indiscriminately, failing to notice that as a result it was underestimating even the 100 per cent cereal basket.[6] The error is fundamental and carried over to the calculations of the full diet: the cost entered against this diet would only be able to buy a 100 per cent cereal basket in the urban areas and not even that in the rural areas. A correctly specified 60:40 cereal/non-cereal diet would itself cost 50 per cent more than the total poverty line in rural areas and just about the same as the total poverty line in urban areas.

Thus the following conclusions are reached about the 1966 poverty line: (1) it was derived from the 1975 line by applying the cost of living index globally; and as a result (2) the poverty lines chosen were much too low, especially in the rural areas.

The Tunisian estimates of poverty lines

The Tunisian authorities used to work with a normative income level below which 'decent' life is not considered possible. After the World Bank published its estimates, this poverty line was partially abandoned in favour of the Bank figures, although in most subsequent Plan documents the normative poverty line was still mentioned. Thus, in effect two poverty lines exist in Tunisia at the moment: one used in the plans and the other in the INS budget surveys. In fact, in the VIth Development

Plan (1982–86) both poverty lines are mentioned and both sets of poverty figures are also shown. We will look first at the Plan poverty line and then at that of the INS.

The Plan lines

The Plan poverty line was established around 1960. It was a national poverty line, as no separate estimates were made for rural and urban areas. Moreover, no details of the methodology are available. According to this estimate, as much as three-quarters of the population were counted as poor in the late 1950s. Over the years this poverty line was updated straightforwardly by applying the cost of living index. At a later stage, it was arbitrarily modified to reflect social benefits provided by the state. By this time the World Bank's poverty line had also become available, and as a result different – sometimes contradictory – estimates of poverty have been advanced for the same year. The extent of these contradictions may be gleaned from Table C.3 and are highlighted below.

1 Each plan document gives a different date for the time that the poverty line was established – 1957, 1961 or 1962; Plan IV even mentions two dates simultaneously.
2 If the poverty line reached D70 in 1968 from D50 in 1957 (Plan II), how did it also reach D70 in 1972 (Plan IV)? Was there no inflation between 1968 and 1972? (There was: 14 per cent or so.)
3 Why is D50 mentioned (implicitly) as the poverty line in 1968 (Plan IV), 1966 and 1975 (Plan V)?
4 If in 1957 75 per cent of the population had an income below D50, how could 73 per cent have less than D43 in 1961?
5 How was it decided to reduce the poverty line to take account of social benefits? How was it decided to value these at D7 in 1961 when in 1971 they were thought to amount to D20? At the inflation rate used they should have been valued at D13 in 1961.

The idea of accounting for social benefits is of course sound, but apart from the peremptory way in which this was done, it can also be faulted conceptually. A properly estimated poverty line should make allowance for the necessary expenditure on health, education and so on. A consumption survey only records consumption actually incurred by the household, whereas a household also consumes social services provided free (or at subsidized rates) by the government. These should be included in a household's total consumption. Thus instead of lowering the poverty line, consumption should be raised.

Table C.3 Poverty lines (in dinars per capita per annum) and poverty (per cent) in four Development Plans

Plan III, 1969–72			Plan IV, 1973–76			Plan V, 1977–81		Plan VI, 1982–86		
1957	D50		1957	D50	75% (p. 161)	*Population earning < D50*		1961	D50 (D43 with social benefits included)	73%
			1962	D50	75% (p. 26)	1966	27%	1966	(D55)	48%
1968	D70 (63%)		1968	D50	43% (p. 162)			1971	n.a.	42%
			1972	D70	40% (pp. 26, 162)	1975	16%	1975	D70	29%
								1980	D100	21%
								Alternative figures		
									(D59)	22%
									(D75)	13%

Remarks

Plan III: Mentions (p. 8) that poverty line was set at D50 in 1957; at 1968 prices, the poverty line was projected as D70.

Plan IV: The 1962 figures must be an error. The 1968 figure is literally correct – 43% had < D50 – but this is misleading since D50 was not the poverty line in 1968. Mentions (p. 162) that monetary incomes do not include social benefits (health, education) put at D20 p.c.p.a. in 1972.

Plan V: Only figures available (p. 10.24), but misleading, as D50 was not the poverty line in 1966 or 1975. 1966 figure also literally incorrect: should read 27% had < D50 at 1975 prices (= D35 at 1966 prices).

Plan VI: Main-line figures from p. 3; alternative figures from p. 212. The latter are INS poverty line/poverty figures based on World Bank methodology. All main-line poverty estimates based on 'social-benefit poverty line'.

Sources: Tunisia Plan documents, years as indicated above.

What about the appropriateness of the poverty line chosen? We cannot say, because we do not know its contents. By comparing it, however, with our line in 1975 (see below) it would appear to be of the right order of magnitude.

The INS lines

The INS first calculated a poverty line for 1980. This was projected backward to 1975 (and later forward to 1985) and the two poverty lines were set against the appropriate household budget surveys to calculate the incidence of poverty. (The Plan documents used the INS budget surveys against their own poverty lines to calculate the incidence of poverty.)

The INS's poverty line came straight from the World Bank, as clearly mentioned in INS documents. Thus, in the 1980 budget survey, in which the poverty line was first proposed, it was stated:

[La] population, considérée défavorisée, vit avec un revenu par tête en deçà d'un seuil bien arrêté qu'est le seuil de pauvreté. Ce seuil, dont les techniques et les méthodologies de calcul sont suggérées par la Banque Mondiale, est considéré comme norme de pauvreté absolue, de telle sorte que les couches de population ayant un niveau de revenu inférieur à ce seuil ont des risques énormes de vivre dans des conditions de pauvreté absolue (p. 69). [The population, considered as disadvantaged, lives with an income per capita below that considered as the poverty line. This line, the techniques and methodology of which were suggested by the World Bank, is considered as the level of absolute poverty, in the sense that the population groups with an income inferior to such a line run a great risk of living in absolute poverty.]

The question we will try to answer can be posed as follows: given that the World Bank methodology put the poverty line at the twentieth-percentile level and the 1980 line was not put at the twentieth-percentile level, how did the INS derive its poverty line in 1980? What seems to have been done is that the INS put the poverty line at 'un revenu oscillant autour de 20ème percentile'.[7] This of course reduces even further the seeming precision in the World Bank methodology. The INS made another arbitrary change from standard practice: calorific requirement was cut from the norm of 2,200 calories per capita per day, to 1,830 calories for rural areas and 1,870 for urban areas.[8]

Given the arbitrary cut-off point chosen (even more arbitrary than that of the World Bank), we should not be surprised that the poverty lines selected do not provide sufficiently for all basic needs. As before, the underestimation is most pronounced for the rural areas. In 1980 a

60:40 cereal/non-cereal diet itself would have almost exhausted the total poverty line income stipulated for the rural areas. The 1980 poverty lines were projected forward to 1985 and backward to 1966.

The following conclusions emerge about the national estimates of poverty lines in Tunisia. First, the plan document poverty line was originally established as an aspirational figure. Over the years many arbitrary changes were made to it. Second, the INS poverty line derived for 1980 was based on the World Bank's modifications to methodology proposed for 1975. The cost-of-living index was applied to arrive at the poverty line in 1966 and 1985. The rural poverty line is low in an absolute sense.

The ILO poverty line

What is needed to set a poverty line are standards showing the minimum amount of food, clothing, shelter and so on necessary to support a 'decent life'. Such standards do not exist. There is general agreement on the minimum calories necessary for ordinary living – 2,200 calories per capita per day – but those 2,200 calories may come from any combination of foodstuffs, from starchy foods to meat. A diet consisting of 80:20 cereal/meat may cost only one-third as much as another in which the proportions are reversed. The same point can be made concerning clothing, shelter, or hygiene, where no standards at all are available.

In this section we shall show: (1) our procedures for estimating the food poverty line in detail, particularly for 1975 to allow a comparison with the World Bank's 1975 poverty line; and (2) the basis for our non-food estimates on various different assumptions.

Food poverty line

Table C.4 shows the calculation of the minimum food basket for 1975. In this basket, 60 per cent of the calories are derived from cereals, 18 per cent from oil, 11 per cent from vegetables, 7 per cent from sugar and 4 per cent from meat. The basket is set to provide 2,200 calories per person per day – the FAO standard. It is a sort of 'average' basket, one which we find from the relevant household budget survey is likely to be consumed at the thirtieth-percentile level.[9] It is the type of basket that could be advanced for a medium-income developing country. Later we shall do a sensitivity analysis to show how costs vary as we change our assumptions about the diet. As for the rest of the figures in the Table, columns 2–4 are data; columns 5 and 6 are manipulations on the data; while columns 7 and 8 are results based on all previous figures.

Table C.4 Composition and cost of a minimum food basket, 1975

	Calorie composition (%) (1)	Calories per kg (2)	Prices (m/kg) Urban (3)	Prices (m/kg) Rural (4)	Price/1,000 calories (m) Urban (5)	Price/1,000 calories (m) Rural (6)	Cost (m/day) Urban (7)	Cost (m/day) Rural (8)
Cereals	60	3,200	90	61[a]	28.1	19.1	37.1	25.2
Oil	18	8,840	358	354	40.5	40.0	16.0	15.8
Vegetables	11	500	96	89	192.0	178.0	46.5	43.1
Sugar	7	3,960	211	209	53.2	52.8	8.2	8.1
Meat	4	1,860	789	683	424.2	367.2	37.3	32.3
Total	100	(= 2,200 calories per person per day)					145.1	124.5

Source: Compiled by the authors.
Note: [a] The price differential between rural and urban areas arises from processing.

Our diet would cost D53 per person per year in urban areas and D45 in rural areas, based on prices as recorded in the household budget survey. Let us compare the World Bank poverty lines with our results (in dinars per person per year):

	Urban	Rural
ILO food	53	45
World Bank food	41	23
World Bank total	79	38

In a very compact fashion the above figures reinforce all the points made earlier about the World Bank estimates: (1) the urban poverty line was of a lower quality than the one chosen here; (2) the rural poverty line was of an even lower quality; therefore (3) the rural and urban poverty lines were not equal; and (4) the Bank's rural basket would not buy even basic foods.

Table C.5 shows a sensitivity analysis as a basis for judging the appropriateness of our poverty line. Rural prices are used throughout. Diet B costs only half as much as diet A, while diet C costs twice as much; the gap between B and C is of the order of 4.3:1. Poverty lines should be 'reasonable', given the context of the country: they should not be set too high by including too many expensive foods; neither should they be set too low by including only the staple foods. As cereals (and other starchy foods, root crops and tubers, not important in Tunisia) are invariably the cheapest sources of calories, setting a 'reasonable' diet always means basing it around cereals; however, there is always the danger of loading it so much with cereals that it becomes unpalatable.

Table C.5 Cost of three different diets at 1975 prices

	Calorie composition		
	Diet A (chosen)	Diet B	Diet C
Cereals	60	90	35
Oil	18	5	20
Vegetables	11	5	15
Sugar	7	–	10
Meat	4	–	20
Cost (dinars per person per year)	45.40	22.56	96.43

Source: Calculated by the authors by applying rural prices in Table C.4 and assuming 2,200 calories per person per day.

Diet B would be such a diet. It is in fact the sort of diet the World Bank ended up with by working backward from their twentieth-percentile formula. Diet C, on the other hand, could be afforded by only the top quartile of the population. (Choosing it would put 75 per cent of the population in 'poverty'.) Thus, our choice (diet A) seems reasonable: it is varied enough, it reflects local eating habits and it is not too far out of reach of the majority of the population.

Table C.6 shows the prices of the food items in 1966 and 1980 and the food poverty lines based on them, the composition of the food basket being maintained at the same pattern as in Table C.4.

Two points may be made about the figures in the Table, which have a bearing on the procedures applied in the derivation of poverty lines in the past. First, as noted before, the price differential between rural

Table C.6 Prices in 1966 and 1980 (millimes/kg)

	1966		1980	
	Urban	Rural	Urban	Rural
Cereals	108	73	125	88
Oil	236	236	500	449
Vegetables	49	44	165	131
Sugar	100	95	343	259
Meat	433	382	1,244	1,390
Cost of food basket (dinars per person per day)	38.58	31.24	83.69	72.31
Differential	1.23		1.16	

Source: Calculated from relevant household budget surveys.

and urban areas is much smaller than the differential used in the World Bank and later poverty lines. The cost differential in the Bank's calculation comes from a quality differential, that is, the urban basket included more expensive foods than the rural basket. Second, the price increase implied by the above figures differs from the price increase in the cost-of-living index, and the price increase between rural and urban areas also differs. Thus, for the urban areas, between 1966 and 1975 the price increase from the surveys was 37 per cent whereas the cost of living increased by 43 per cent. Between 1975 and 1980 the corresponding figures were 58 per cent and 40 per cent. As mentioned earlier, the World Bank and the INS took the 1975 poverty line and extrapolated the 1966 and 1980 poverty lines by applying the cost-of-living index. Given the differences in the observed rates of inflation, in both years the resulting figures would give a lower poverty line than warranted, the difference in 1980 being significant. Thus, the suggested methodology is not very convincing: the 1975 poverty lines were placed too low and the methodology applied resulted in even lower poverty lines in the other years.

Non-food poverty line

Problems in establishing equivalent poverty lines for non-food components are legion:

1 No standards exist for non-food needs, although some could be devised, for example, metres of clothing, cubic metres of housing space, kilograms of cleaning materials and so on.
2 Price data for rural and urban areas are not readily available.
3 The quality of non-food goods differs between rural and urban areas, for example, in housing. This could be partially captured by the price differential.
4 Needs differ between rural and urban areas, for example, in transport, clothing and medical care.

Given sufficient data, it is just conceivable that approximately equivalent poverty lines could be constructed. Such data are not available and hence we have resorted to a simple estimating procedure to arrive at the non-food basket. This procedure is based on a detailed examination of 1975 household budget survey figures. The non-food poverty line for urban areas was put at 40 per cent of the average non-food expenditure; the rural basket was then put at 55 per cent of the urban basket. The resulting poverty lines are shown in Table C.7, along with figures of average expenditure from the various household budget surveys.

Table C.7 Average expenditure from household budget surveys and estimated poverty lines (dinars per person per year)

		1966		1975		1980	
		Urban	Rural	Urban	Rural	Urban	Rural
1	Non-food expenditure	72.3	22.7	119.0	55.5	166.5	79.4
2	Food expenditure	52.3	30.4	73.5	50.3	117.5	77.3
3	Total expenditure	124.6	53.1	192.4	105.8	284.0	156.7
Poverty lines							
Non-food (urban = 0.4 of							
(1); rural = 0.55 of urban)		29	16	48	26	67	37
Food		39	31	53	45	84	72
Total		68	47	101	71	151	109

Sources: Household budget surveys previously cited. Figures for 1966 from pp. 258, 259, urban = *grandes villes* and rural = *dispersé*; 1975 from p. 228, urban = *grandes villes + milieu urbain*; 1980 from vol. 2, p. 85, urban = *communes urbaines*.

1985 poverty line

For 1985 we have adopted a somewhat different procedure from the above in order to investigate further the range in the rural–urban cost of living. We have taken prices of most foodstuffs to be the same in rural and urban areas, as is the case in Tunisia because of the *prix homologues* system of pricing. Two exceptions are vegetables and meat, where rural prices are set at three-quarters of urban prices. We then set different cereal baskets for rural and urban areas, keeping the non-cereal basket the same, the latter on the ground that the diet chosen is already at a rock-bottom level and therefore does not allow for further cuts in non-cereal consumption. In the case of cereals, while staying within a 60 per cent cereal diet, we gave the rural population the option of changing their diet to cheaper sources, for example, semolina instead of the prepared couscous, farina instead of bread. The results are shown in Table C.8. The most important lesson from this exercise is that there is very little scope for savings on cereals through substitution. Thus, the more labour-intensive rural cereal basket costs only 23 per cent less than the urban basket. (The reason for this, of course, is that the price regime in Tunisia favours the more processed cereals through subsidies – an aspect favouring the urban areas, as discussed in the text.) The price differential for vegetables and meat similarly is quite limited. Thus, altogether the rural food basket is some 21 per cent less costly than the urban basket. On a yearly basis the rural food basket would cost D107 per capita, and the urban basket D136. In comparison, it is worth mentioning that the INS's food basket for rural areas was only around two-thirds the cost of the urban basket.

Table C.8 Cost of a food basket in 1985

	Diet composition (%)		Price (millimes/kg)	Price per 1,000 calories	Cost (millems per day)	
	Urban (u)	Rural (r)			Urban	Rural
Semolina	0	25	155	55	0.0	30.3
Couscous	15	0	230	72	23.8	0.0
Macaroni	15	10	225	75	24.8	16.5
Farina	0	25	115	36	0.0	19.8
Bread	30	0	142	57	37.6	0.0
Oil	18	18	530	60	23.8	23.8
Vegetables	11	11	200(u) 150(r)	400 300	96.8	72.6
Sugar	7	7	61	154	23.7	23.7
Meat	4	4	3,000(u) 2,250(r)	1,673(u) 1,210(r)	141.9	106.4
Total	2,200 calories	2,200 calories	–	–	372.4 135.93	293.1 106.98

Source: Compiled by the authors.

Notes

Preface

1 Bedoui Abdel-Jelil, 'L'Emploi Non-Agricole et Urbain en Tunisie';
   Mohamed Ayyad, 'Evolution of Rural Development Programmes since
   Indepenence' (in Arabic); Messaoud Boudiaf, 'Desertification de la
   Campagne Tunisienne'; Hussein Dimacy, 'Patterns of Employment and
   Income in the Tunisian Village'; Omar Kaddour, 'Scenarios de
   Projection de la Main d'Oeuvre Agricole'; Khaled Louhichi, 'Internal
   Migration and Rural Development in Tunisia' (in Arabic); and Khalil
   Zamiti, 'Les Effets des Transformations Economiques et Sociales sur la
   Famille Rurale'.

1 Introduction

1 There is a direct relationship between income and life expectancy.
   Thus, based on country cross-section data, the World Bank (in
   World Development Report, 1980, p. 36) shows an upward-sloping
   curve of life expectancy on income. At an income of around US$1,100
   (the Tunisian average in 1986) the life expectancy norm was 60
   years. Tunisia exceeded this. Most oil-producing countries fall below
   the norm as may be ascertained by comparing actual life expectancy
   (from *World Development Report, 1988*, p. 286) against the norm.

2 Employment and labour markets

1 This is an indication of increasing use of contraceptives by married
   women of childbearing age. Thus, according to World Bank data, the
   percentage of such women using contraceptives increased from 10 per
   cent in 1970 to 42 per cent in 1985: World Bank, *World Development
   Report, 1988*, Table 28. (Figures of birth and death rates cited
   previously are from *World Development Report*, 1986, Table 26.)
2 For an excellent discussion of the structural transformation process see,
   for instance, Bruce F. Johnston and William C. Clark, *Redesigning
   Rural Development: A Strategic Perspective*, Baltimore, Johns Hopkins
   University Press, 1982, pp. 38–44.

3 There is a vast literature on rural exodus. For a good summary see Khaled Elmanoubi, *Internal Migration in Tunisia* (Population Research Unit, League of Arab States, Tunis, March 1986). For an analysis of the rural–urban migration waves of the 1960s and 1970s, see Khaled Louhechi, *Rural Development and International Migration in Tunisia, with Special Reference to the North-west*, Tunis, Population Research Unit, League of Arab States, December 1985.

4 Jacques Charmes, *Secteur non structuré: Politique, économique et structuration sociale en Tunisie, 1970–1985* (mimeo), 1985, p. 4.

5 See, for example, Abdel-Jelil: *L'emploi non-agricole et urbain en Tunisie*, Tunis, League of Arab States, 1986.

6 Jacques Charmes, 'Deux estimations de l'emploi dans le secteur non structuré en Tunisie: résultats de l'analyse comparative', in *Seminaire sur les statistiques de l'emploi et du secteur non structuré*, Paris, July 1985, vol. 2, Annexe méthodologique, pp. 450–64.

7 For example, with the informal sector accounting for 50 per cent of the urban labour force and the modern sector 20 per cent, if all of the labour force growth of 4 per cent per annum has to be absorbed in the modern sector, it would have to grow by 20 per cent per annum, whereas the equivalent growth in the informal sector would have to be 8 per cent.

8 Jacques Charmes, *Secteur non structuré, op. cit.*, p. 4.

9 A quotation from A. O'Connor, *The African City*, London, Hutchinson 1983, p. 141, sums up our argument:

> [I]n drawing a distinction between large-scale enterprises . . . and smaller-scale local enterprise, we must recognize that in some cities the latter largely pre-dates the colonial intrusion whereas in others it has followed it. There are fundamental differences between pre-colonial cities (or parts of cities), where petty trade and crafts form the historic basis of the urban economy and where the 'formal sector' may even remain peripheral, and those cities where the latter developed first and an 'informal sector' has emerged dependent upon it.

Tunisia falls in the former case. North African Arabs themselves make a distinction between *souk al Arabi* (the Arab market) and *souk farenji* (foreign market). The former consists of the myriad of small-scale activities – commercial as well as industrial – that are conducted in the alleys and by-ways of the old town. These have existed from time immemorial. *Souk farenji*, on the other hand, is the modern sector. For a vivid picture of Tunisian artisanat, see Fouad Ibrahim, *Das Handwerk in Tunisien*, Hanover, 1975.

10 Jacques Charmes, *Secteur non structuré, op. cit.*, Table 3, p. 13.

11 Ibid., p. 15.

12 Ibid., Table 4, p. 16.

13 In agricultural and food industries the ratio of a qualified worker's wage to the SMIG was 1.7 and for semi-qualified workers it was 1.3. Charmes, ibid., Table 5, p. 18.

14 Ibid.

15 Hussein Dimacy, *Patterns of Employment and Incomes in the Tunisian Village*, Tunis, League of Arab States, 1986, p. 20.
16 These rates range from 2 to 10 per cent in developing countries. See ILO, *World Labour Report I*, Geneva, 1984. As will be noted in Appendix B, our estimates are higher than official ones, since the latter define the unemployed as those who are out of work in the age group 18–59. This leaves out the unemployed in the 15–17 age group (as well as those over 60). Since unemployment among the latter groups is extremely high, their inclusion raises the unemployment rate from 12.9 and 11.5 to 15.7 and 16.4 in 1975 and 1984 respectively.
17 INS, *Enquête population – emploi, 1980*, p. 128.
18 Ministère des Affaires Sociales, Office de la Promotion de l'Emploi et des Travailleurs Tunisiens à l'Etranger, *Le chomage en Tunisie: Niveau, origine et structure. Résultats de l'enquête sur l'emploi et le chomage de 1983*, Tunis, April 1984,
19 INS, *Recensement général de la population et de l'habitat, 1984*, vol. 5, p. 51.
20 Ibid., pp. 12 and 103.
21 INS, *Recensement général de la population, 1975*, vol. 5, p. 49.

4 Income distribution and poverty

1 It is worth mentioning that data in the HBSs are generally not given so as to enable one to calculate the Gini coefficients. Thus, data are given in the form of Table 4.5, without additional columns showing the percentage of consumption that went to each population group.
2 In this respect it may be noted that surveys done by the Ministry of Social Affairs in 1986–87 as part of the programme, 'Aide pour les familles nécessiteuses', established a list of 125,000 needy families based on the personal knowledge of local social workers. Most of these families were handicapped in some real sense – age, widowhood, disability, etc. – and had practically no source of income. Most of the 125,000 families happen to be in the rural areas. Thus, the Ministry could identify over 100,000 needy families in the rural areas simply on the basis of physical handicap. In contrast, the INS arrived at a figure of only 40,000 families in poverty based on a criterion of income. The World Bank has now recognized the error in these earlier estimates of the poverty line, while the INS at the time of writing (1988) had commenced a review of its estimation procedures.

5 Food consumption, food balances and subsidies

1 The authoritative source on food consumption patterns remains M.K. Bennett, *The World's Food*, Salem, New York, Ayre Co., 1954.
2 Figures based on data available in the first source cited for Table 5.5.
3 World Bank, *Tunisia: Social Aspects of Development*, Washington, DC, June 1980, Table 24.

Appendix A: Labour migration from Tunisia

1 Ministère des Affaires Sociales, Office de la Promotion de l'Emploi et des Travailleurs Tunisiens à l'Etranger, *Rapport du Comité Technique de l'Emigration*, Tunis, June 1985.

2 INS, *Recensement général de la population et de l'habitat, 1984*, vol. 5.

Appendix B: Estimates of unemployment

1 The census states that 'pour les jeunes de 15 à 17 ans on peut dire qu'ils relèvent plutôt du domaine de la formation, d'autant plus que les créations d'emploi dans les secteurs modernes de l'activité ne sont pas accessibles aux personnes âgées de moins de 18 ans', see INS, *Recensement général de la population et de l'habitat*, 1984, vol. 5, p. 11. [It could be said that youth aged 15–17 belong rather in the training category, particularly as the under-18 age group is not eligible for employment in the modern sector.]

2 The census justifies this thus: 'Pour les personnes âgées de plus de 60 ans, on peut dire qu'elles ont atteint un âge déjà assez avancé et ont besoin d'une assistance sociale plutôt que d'exercer un emploi', ibid., p. 11. ['Persons aged 60 and over have reached a relatively advanced age and are more in need of social assistance than employment.']

3 For this and the next category in (c), the census of 1984 adopted the following procedure:

En plus des personnes ayant déclaré exercer une activité la semaine précédant le recensement, nous avons considéré comme 'actives occupées' les personnes appartenant aux deux catégories suivantes:

Certains ménages ruraux, disposant d'une exploitation agricole ont déclaré que tous les membres du ménage sont en chômage ou inactifs, dans ce cas le chef du ménage a été considéré actif occupé chargé de l'exploitation agricole, et classé en qualité d'exploitant agricole.

Certaines femmes n'ont pas considéré leur activité dans l'artisanat ou leur participation aux travaux agricoles comme une activité économique et se sont déclarées spontanément 'femme au foyer' sous prétexte que le marché de l'emploi ne peut les intégrer; ces femmes, surtout celles âgées entre 18 et 59 ans, ont été assimilées à des actives occupées en tant qu'indépendantes à domicile ou aides familiales'. (INS, *Recensement général de la population et de l'habitat, 1984*, vol. 5, pp. 10–11. ['In addition to persons having declared themselves as employed during the week preceding the census we counted as ''gainfully employed'' the following two categories:

Some rural households with a farm declared all their members as unemployed or not in activity. In those cases the household head has been considered as employed in charge of the farm and counted as a farmer.

Some women did not consider their small-scale crafts or agricultural work as an economic activity and declared themselves spontaneously as housewives with the assumption that the labour market could not integrate them. Such women, particularly those in the 18–59 age group, were counted in gainful employment as independent homebased workers or family workers.']

Appendix C: Poverty lines in Tunisia: A critical review

1 For details of these estimates see World Bank, *Tunisia: Social Aspects of Development*, Washington, DC, June 18, 1980, pp. 9–14. It is worth noting that this report has not been widely circulated in Tunisia; many believe it is confidential.
2 INS, *Enquête nationale sur le budget et la consommation des ménages, 1975*, Tunis, 1978.
3 Ibid., pp. 233 and 397. The following figures summarize the situation for the urban areas:

	Expenditure (millimes per day)	Calories	Millimes per 1,000 calories
Cereals	38	1,223	31
Non-cereals	163	1,201	136
Total	201	2,424	83

4 It may be noted that this is the average urban diet according to the 1975 budget survey (p. 397).
5 Rural to urban ratios: line 1 and 3, 0.68; line 2, 0.57; line 4, 0.45; line 5, 0.38; line 6, 0.47. Food expenditure as a percentage of poverty line: urban 52, rural 74 in both years. How the food expenditure of the twentieth-percentile group is obtained in each year is not clear.
6 The error arises from the fact that cereal prices did not move in line with the cost of living index: in fact, cereal prices were higher in 1966 than 1975. The following prices are obtained from the two budget surveys (millimes/kg):

	Rural	Urban	Calories/kg
1966	73	107	3,500
1975	61	82	2,900

In terms of calories, prices were about equal in the two years.
7 H. Farouti, 'Pourquoi un seuil de pauvreté?', INS, 1986, paper presented at an FAO conference on poverty estimates.
8 Ibid.
9 The point noted earlier about the futility of trying to derive a poverty line from a household budget survey may be recalled. We are not attempting to estimate the poverty from the budget survey but simply making a comparison of an independently derived poverty line with figures from the household budget survey to judge its order of magnitude.

Index

Abdel-Jalil, B. 18, 19, 59, 113
agriculture 29–44; balance of
  payments 74–5; co-
  operativization *see* co-
  operativization; development
  strategy 90–1; GDP 29, 50–1;
  investment 2, 3, 29, 30, 32,
  39–41; labour force 4–5, 10,
  12–17, 26, 29, 50; landholding
  36–9; policy 29, 30–1, 35–6,
  41, 44; prices and subsidies
  41–4; productivity *see*
  productivity; subsidies 29, 31,
  36, 43–4, 88; technological
  change 35–6, 43; wages 50–1,
  114
aid, foreign 30, 44, 91
Algerians 3, 92
Arab countries 92, 93

bakery bread 67, 71, 78; *see also*
  bread
barley 44, 77, 78, 82; inferiority
  69–71; shift to 32–3; subsidies
  83
benefits, social 104
Bennett, M.K. 115
birth rate 10
boom 3–4, 88, 89
Boudhiaf, M. 33, 40
bread 67, 77, 78, 82; convenience
  71; subsidies 72, 84, 85

calories: cereals and 63–4, 77–8,
  82, 101–2, 117; daily supply per

capita 5, 6; poverty lines and
  101–2, 106, 107, 117
capitalism 31, 32, 36, 41, 44
casualization of labour 13–16
cereals: consumption 63–6; imports
  75–82, 87; landholding and 37,
  38; poverty line 101–2, 102–3,
  107, 108–9, 111, 117;
  preferences 66–73; prices 42,
  71–3, 100, 102–3, 117;
  production 76–7, 78, 79;
  subsidies 43–4, 82–4
Charmes, J. 18, 18–19, 113, 114
circular migration 14
Clark, W.C. 113
class, social 47, 50
clothing/textile industry 3, 20, 21,
  22, 23
coastal areas 11, 52, 88
construction industry 11
consumption: dilemma 67; food
  63–6; income distribution and
  45–50, 62; poverty and 56, 104;
  wheat preferences and 66–7, 69,
  70, 72, 76–7
co-operativization 31, 35, 44, 92;
  policy 3, 30; productivity and
  32, 34
cost of living index 102, 103, 104,
  107, 110, 117
couscous 66–7, 71, 72, 77
credit, agricultural 36, 40–1, 43,
  88
crisis, economic 2, 89
crop production 32–4, 36, 38, 91

dates 76
debt, public 2, 6, 90
decolonization 3
deficits, trade 74-5
demand, inability to meet 61, 82, 87
development; new forms of production and 23; strategy for 90-1
Development Plans 103-6, 107: Sixth 2, 24, 103-4, 105; Seventh 2, 89-90
Dimacy, H. 114
duration of unemployment 26-7

economic crisis 2, 89
economy: structural imbalances 90; transformation of 1-6, 10, 61, 62, 89
education 27
Egypt 85
Elmanoubi, K. 113
emigration 3, 4, 23-5, 92-4; *see also* migration of labour
employment: decolonization 3; development strategy 89, 91; migration and 94; shifts in patterns 4-6, 87; structural transformation 4-6; *see also* labour force; unemployment
Enquête Population - Emploi 51
equity: growth and 61; subsidies and 86
Europe 17, 23, 25, 92, 93; *see also* France
expenditure: food 63-4, 99, 101, 102, 117; household budget surveys and 48-9
exports 29, 75, 76; promotion 89-90
external resources *see* petroleum; remittances; tourism

families, poverty and 60, 115
family labour 13-14, 16, 21-2
farina 111
farmers, poverty and 55
Farouti, H. 117
female labour force 26, 95-7, 116

fertilizers 32, 35, 36, 38, 43
fish 42, 43, 44
food 31, 61, 63-86, 87, 90; balances 29, 63, 74-82; cereal preferences 66-73; consumption 63-6; industries 114; poverty line and 53-4, 99-103, 107-10, 111-12, 117; subsidies 72, 82-5, 85-6, 111
Food and Agriculture Organization (FAO) 52, 77, 99, 107
forage crops 33, 37, 38, 44
foreign aid 30, 44, 91
foreign exchange 2, 3, 4, 88, 89
foreign market 114
forestry 41
France 3, 4, 23, 92, 93
fruit 41, 64, 76, 77, 78; prices 42, 43, 100; subsidies 44

Gafsa 26
Gini coefficient 48, 49, 50, 62
Griffith-Jones, S. 84
gross domestic product (GDP): agriculture and 29, 50-1; external resources and 4; growth 4, 5, 87; stagnation 6; subsidies and 83-4; wages and 50-1, 89
growth 1, 52, 62, 87-9; equity and 61; indicators 4-6; population 8-10; trickle-down 56, 61, 89

hard wheat 66-73 *passim*, 77, 83, 85
Harvey C. 84
high-yielding seeds 32, 35, 44
home-based work 20-1, 95-7, 116
household budget surveys (HBS) 52, 63, 84, 115; consumption 45, 48-9, 56; food expenditure 101; poverty and 56, 102, 107, 117
households, size of 46
housewives 95-7, 116
housing 101

Ibrahim, F. 114
imports 36; food 74-9 *passim*, 82, 87
income: food consumption and

64–6, 67, 68, 69, 82; informal
sector and 21; life expectancy
and 113; per capita 1, 6, 87, 89,
102; subsidies and 83–5, 86; *see
also* wages
income distribution 45–9, 56, 61,
62, 89 trends 49–52
industrial crops 42–3, 43–4
industrialization, subsidized 2, 3–4, 88
infant mortality rate 5, 6, 8
informal sector 17–23, 88, 114
Institut National de la Statistique
(INS) 45, 48, 117; poverty
53–4, 57, 60, 62, 98, 106–7,
115 (cost of living index 56,
107, 110; rural 53, 107, 112);
subsidies 83–4
International Labour Organization
(ILO) 54, 57, 107–12, 114
International Monetary Fund 2
investment: agriculture 2, 3, 29, 30,
32, 39–41; strategy 90
irrigation 33, 34, 38; investment
32, 35, 41, 43

Jendouba 26
Johnston, B.F. 113

Kassérine 26

labour force 8–28, 96; agricultural
4–5, 10, 11–17, 27, 29, 50;
informal sector 17–23; migration
see migration of labour;
structural transformation 4–6,
31; trends 8–11; *see also*
employment; unemployment
land: decolonization and 3;
ownership and unemployment
25–6, 95, 116; state policy and
property rights 30 (development
strategy 90–1)
landholding 36–9; labour and
14–16, 17
Le Kef 26
Libyan Arab Jamahiriya: emigration
to 4, 17, 23, 88, 92, 93;
expulsions from 25
Libyans 3, 92

life expectancy 5, 6, 113
livestock farming 33–4, 35–6, 3–9,
42, 91; subsidies 43–4
localization, informal sector and 18,
20–1
Louhechi, K. 113

Maghreb 22
maize 71, 82, 83
manufacturing 4, 5, 20, 22
marine products 76
market-determined prices 42, 43
markets, Arab and foreign 114
meat 64, 75, 107; consumption 77,
78, 82; decline in production 33;
prices 100, 111; subsidies 84
mechanization: agriculture 28, 88,
91; casualization of labour
13–14; migration and 11; *see
also* technological change
migration of labour 3, 10–12, 28,
61, 88; agriculture and 11, 14,
31; circular 14; from Tunisia 3,
4, 23–5, 92–4
milk 75
minimum wage 58, 59, 60
Ministry of Social Affairs 23–4,
115
Mishalani, P. 84
mobility, labour 11
modern (private) sector 18, 19, 21,
23, 114
Morocco 85
mortality rate 5, 6, 8

non-traditional exports 90

occupation, poverty and 54–5
O'Connor, A. 114
oil (petroleum) 2, 3, 4, 44, 52, 88
oils, edible 77, 78, 82, 83, 107;
olive 75, 76

pain de boulangerie (bakery bread)
67, 71, 78; *see also* bread
per capita income 1, 6, 87, 89, 102
petroleum 2, 3, 4, 44, 52, 88
Plans, Development *see*
Development Plans

poultry farming 33, 34, 43-4
poverty 26, 45, 52-61, 62, 89, 115;
    incidence 53-6; trends 56-61
poverty lines 45, 52-4, 69, 98-112,
    117; *see also* International
    Labour Organization; Institut
    National de la Statistique; World
    Bank
preferences, cereal 66-73
price equalization fund (PEF) 82-3
prices 50, 76, 85, 117; agricultural
    products 41-4; rural-urban
    differentials 99-103, 109-10,
    111-12; wheat preferences 71-3,
    82; *see also* subsidies
private investment 40-1
production, development and forms
    of 23
productivity: agricultural 6, 11,
    27-8, 41, 87-8 (cereals 76-7,
    78, 79; subsidies 43-4; trends
    31-4); informal sector 21
property relations 30, 90-1
protectionism 88; *see also* subsidies
public investment 40-1
pulses 33, 37, 38; prices and
    subsidies 42, 43, 43-4
'putting-out' system 20-1, 95-7,
    116

*Rapport du Comité Technique de
    l'Emigration* 92
remittances 4, 24-5, 61; growth and
    2, 3, 44, 52, 88
rice 69, 71
Rowdhome, M.B. 37
rural households, unemployment
    and 95, 116
rural poverty 53-4, 56-60 *passim*
    62, 99, 107; remittances and 61;
    see also poverty lines
rural-urban income gap 45-6, 48,
    51
rural-urban migration 10-11, 28
rural-urban price differentials
    99-103, 109-10, 111-12

savings 49
semolina 66-7, 71-2, 111

sensitivity analysis 108-9
SMIG 58, 59, 60
social benefits 104
social class 47, 50
soft wheat 67-73 *passim*, 76-8, 84,
    85
soil conservation 41
sorghum 69-71
souk al Arabi (Arab market) 114
souk farenji (foreign market) 114
soya oil 75
stabilization programme 2
state: agricultural policy 29, 30-1,
    35-6, 41, 44 (development
    strategy 91); external resources
    2-3; price controls 41-3; role in
    development 3-4; *see also*
    investment; subsidies
state farms 30
structural imbalances 90
structural transformation 1-6, 10,
    61, 62, 89
sub-Saharan Africa 19, 22, 71
subsidies: agricultural 29, 31, 36,
    43-4, 88; food 72, 82-5, 85-6,
    111; industrialization 2, 3-4, 88
substitution, food 111
sugar 75, 82, 83, 107

targeting 85, 86
tea 75, 100
technological change 35-6, 43; *see
    also* mechanization
temporary labour 13-16
textile industry 3, 20, 21, 22, 23
time, food preferences and 66, 69
tobacco 75
tourism 2, 3, 4, 11, 52, 88
tractorization 35
trade, agricultural 29, 74-5; *see
    also* exports; imports
transformation of the economy 1-6,
    10, 61, 62, 89
transport 101
trickle-down of growth 56, 61, 89
Tunis 52

underemployment 14
unemployment 2, 11, 25-7, 28, 31,

88; estimates of 95-7, 114, 116; *see
also* employment; labour force
United Nations (UN) *see* Food and
Agriculture Organization
urban poverty 55-6, 56-60 *passim*,
62, 99; *see also* poverty lines
urbanization 10, 87; food
preferences 66, 69, 70; income
distribution 46-7; *see also* rural-
urban income gap; rural-urban
migration; rural-urban price
differentials

Van Ginneken, W. 56, 57
vegetables 64, 107; consumption
77, 78; cultivation 33, 37, 38;
prices 42, 100, 111; subsidies
43-4
wage workers: agriculture 13, 14;
poverty and 55, 60-1

wages 22, 50-1, 89, 114
water *see* irrigation
weather conditions 43
welfare 6, 62, 88-9
wheat 44, 83, 85; preferences
66-73; production and imports
76-9
wine 76
women in the labour force 26,
95-7, 116
wood industry 22
World Bank 2; poverty line
52, 98-103, 104, 109,
117 (comparisons 53-4,
108, cost of living index
56, 110, errors 115, INS and
106, quality differential 110);
subsidies 84

youth unemployment 26, 95, 116